Katrin Wollny-Goerke, Kai Eskildsen (Eds.)

Marine mammals and seabirds in front of offshore wind energy

Katrin Wollny-Goerke, Kai Eskildsen (Eds.)

Marine mammals and seabirds in front of offshore wind energy

MINOS – Marine warm-blooded animals in North and Baltic Seas

Authors:
Dieter Adelung, Sven Adler, Harald Benke, Marie-Anne Blanchet,
Michael Dähne, Volker Dierschke, Jörg Driver, Kai Eskildsen, Stefan Garthe,
Anita Gilles, Nils Guse, Helena Herr, Christopher Honnef, Annette Kilian,
Kristina Lehnert, Paul A. Lepper, Nikolai Liebsch, Klaus Lucke, Nele Markones,
Anja Meding, Gabriele Müller, Tanja Rosenberger, Jacob Rye,
Meike Scheidat, Ursula Siebert, Nicole Sonntag, Janne Sundermeyer,
Ursula Katharina Verfuß, Katrin Wollny-Goerke

Teubner

Bibliographic information published by Die Deutsche Bibliothek
Die Deutsche Bibliothek lists this publication in the Deutsche Nationalbibliographie; detailed bibliographic data is available in the internet at <http://dnb.ddb.de>.

MINOS – Marine warm-blooded animals in North and Baltic Seas

MINOS and MINOS+ are network research projects of prestigious German marine research institutions performed between 2002 and 2008. The administration of the National Park Schleswig-Holstein Wadden Sea manages and coordinates the network.

Nationalpark
Wattenmeer
SCHLESWIG-HOLSTEIN

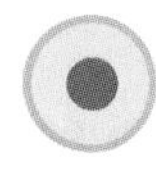

The endeavour on which this publication is based received financing from the German Ministry for Environment, Nature Conservation and Nuclear Safety (BMU) under the funding identification 0329945 (B, C, D). Responsibility for the content rests with the individual authors. The publication was kindly supported by the Jülich Research Centre.

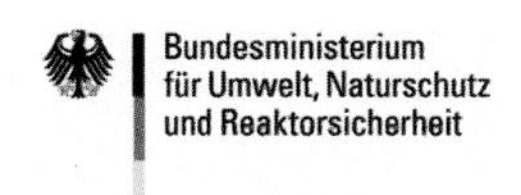

First edition 2008

Editorial Office: Ulrich Sandten / Kerstin Hoffmann

B.G. Teubner Verlag is a company in the specialist publishing group Springer Science+Business Media.
www.teubner.de

Cover design: Ulrike Weigel, www.CorporateDesignGroup.de
Printing and binding: printing company Strauss, Mörlenbach
Printed on acid-free paper
Printed in Germany

ISBN 978-3-8351-0235-4

Content

1 MINOS – the idea behind the research network

Kai Eskildsen

Background – de jure

Winds on oceans are stronger, more consistent and less turbulent than on land, so they can at least in theory produce more energy. Consequently, a discussion started in technical society after the turn of the century whether and where it would be possible and profitable to build wind farms in the Exclusive Economic Zone (EEZ) of Germany. One aspect of feasibility is related to marine legislation. The generation of energy by wind farms in the EEZ requires authorisation from the Federal Maritime and Hydrographic Agency (BSH). The basis for authorisation is the Marine Facilities Ordinance (SeeAnlV). Article 3 emphasises that authorisation must be refused if a marine environment is endangered and cannot be compensated. Specifically, a threat to bird migration is defined there as one aspect of an endangered marine environment.

Figure 1: Wind farm Horn´s Rev, Denmark

The German government also strives to produce energy in an environmentally friendly way by maximising the use of renewable sources. Offshore wind farms will play an important role in the overall concept of Germany's generation of electricity, and corresponding legislation was enacted during the last years. The Renewable Energy Sources Act (EEG) stimulates further planning of wind farms in offshore areas because a guaranteed price for the produced electricity is paid for 20 years.

Subsequently, many applications for projects in the EEZ of the North and the Baltic Seas were submitted to BSH. By November 2007, fifteen wind farms were licensed for the North Sea (Fig. 2), and three wind farms are approved for the Baltic Sea.

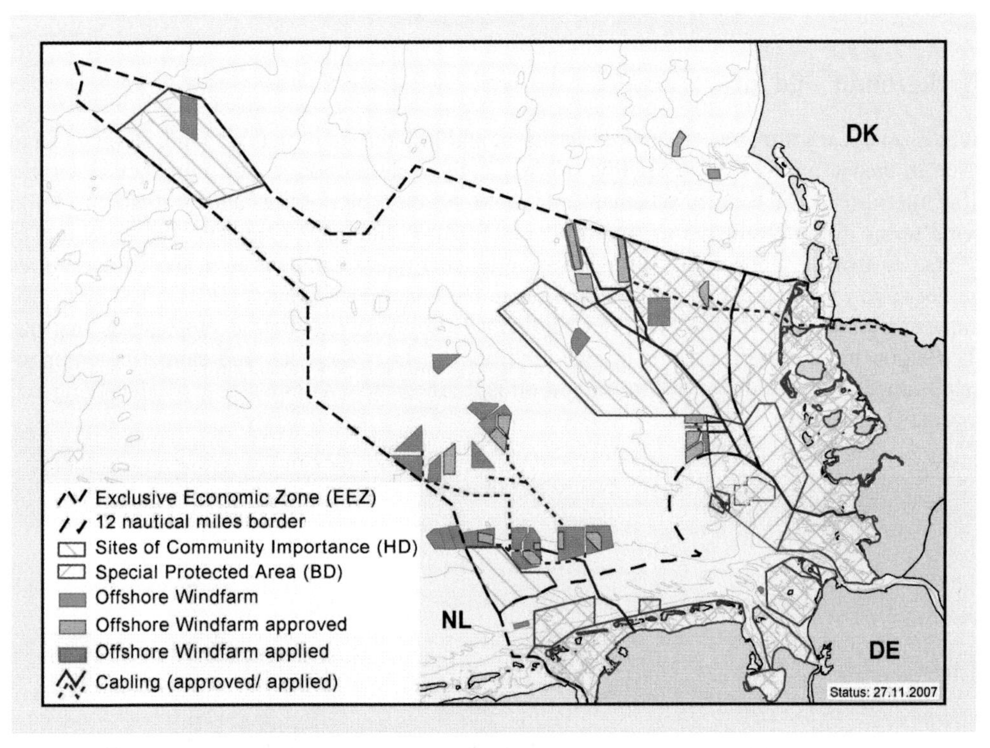

Figure 2: Licensed wind farms (changed after BSH 2007) and protected areas in the framework of Natura 2000 (changed after BfN 2007) in the North Sea. The Natura 2000 network is based on the European Habitat Directive (HD) and European Bird Directive (BD).

Environmentally friendly also means that impacts of the erection of wind farms on the marine environment must be detected and assessed. The expected impacts of offshore wind farms constitute additional threats to animals, adding to existing interferences like fishery, digging for flint or military activities. In this respect it is important to have in mind that simultaneously the Natura 2000 network in the marine environment of the German EEZ was established (Fig. 2). Harbour porpoises, common seals, many seabird species and their habitats are under special protection of the EU Habitat and Bird Directive.

Consequently, marine experts identified different impacts on different scales that may be emerged by offshore wind farms during erection and/or operation:

- Do offshore wind farms affect migration routes of birds, mammals and/or fish?
- Do wind turbine sound emissions during erection and operation harm marine mammals?
- Which ecological relevance do potential sites of wind farms have for marine mammals, birds and their habitats?

- Are direct losses of birds to be feared due to the collision with wind turbines?
- Do electromagnetic fields originating from sea cables impair marine organisms?

Background - scientifically

Some experience with small offshore wind farms situated close to the Baltic coasts of Denmark and Sweden exists but cannot be directly applied to the German seas. The first true offshore wind farm with 80 wind turbines was erected at Horns Rev, Denmark, in the summer months of 2002. Thus, the data available to answer questions arising in connection with the German North and Baltic Seas are rather limited. The three following examples illustrate the dilemma.

Harbour seals (*Phoca vitulina)* are known as one of the most common mammalian predators spread over the North Sea. Since seal hunting was stopped in the early 1970s, harmonised monitoring under the umbrella of the Trilateral Wadden Sea Cooperation was established (Fig. 3).

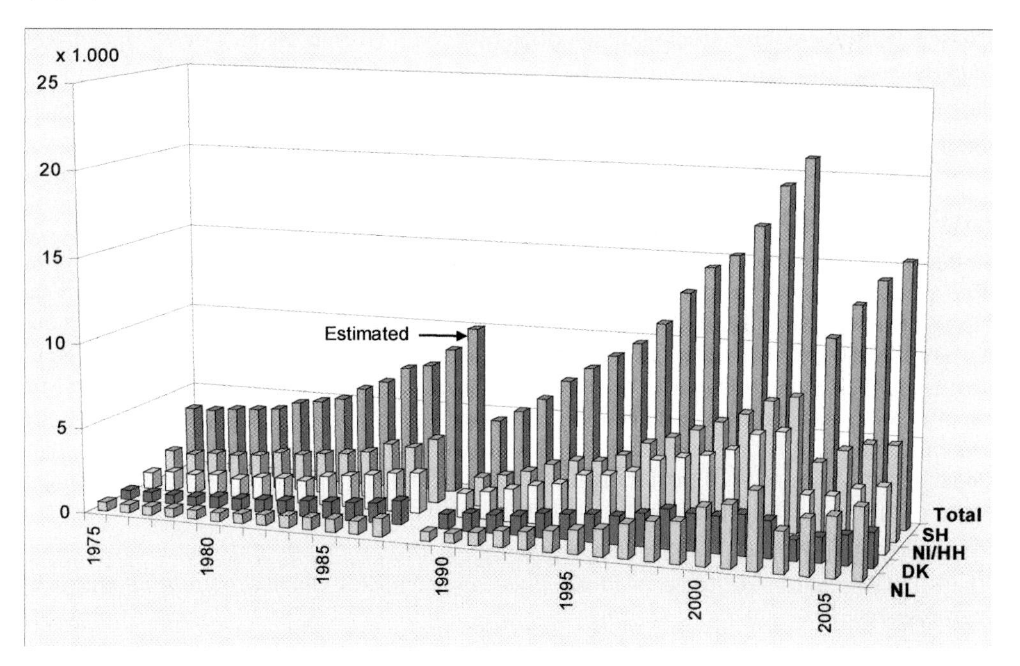

Figure 3: Number of counted harbour seals in the Wadden Sea since 1975 (Common Wadden Sea Secretariat 2006)

The monitoring, however, only takes place on tidal flats where seals rest and moult. Today, knowledge on the population status of harbour seals in the Wadden Sea area is substantial. But what are harbour seals doing when they leave their resting areas? Are they feeding within the area or are they using the adjacent regions? Scientific data to answer these questions are still sparse.

Harbour porpoises (*Phocoena phocoena)* are mammalian predators of the North and Baltic Seas. They are the only cetacean species regularly found in both the German North and Baltic Seas. According to historic reports, harbour porpoises were much more common in coastal waters off Germany in former centuries. However, only little data existed on their distribution in German waters. Most information on distribution and population sizes in the German North and Baltic Seas was based on results of the SCANS project (Small Cetacean Abundance in the North Sea and Adjacent Waters) in the summer of 1994. It generated the first large-scale abundance estimates for harbour porpoises in summer. Unfortunately, some German areas like the waters east of Rügen in the Baltic Sea and coastal waters of Lower Saxony were not included (Hammond et al. 1995). SCANS also did not generate data for the three other seasons. It was nevertheless the only effort on a comparable scale to estimate cetacean abundance in the German North and Baltic Seas before the beginning of MINOS.

Birds, especially seabirds, are of great interest to the public. First attempts to protect special habitats of birds were undertaken in areas at the German coastline (e.g. Kempf et al. 2000). But almost 30 species spend most periods of their life at sea. It is known that species like red-throated diver (*Gavia stellata*), black-throated diver (*Gavia artica*) or Common Scoter (*Melanitta nigra*) are very sensitive to shipping traffic. They also do not breed in Germany. Consequently, these three species are not often observed by laypeople and knowledge of their requirements at sea is limited.

The MINOS research network was the first coordinated large-scale effort to answer these questions related to seabirds and mammals.

Establishing the MINOS network

Under these preconditions the idea of a research network (MINOS and MINOS+) was born. Two projects, MINOS and its successor MINOS+ (Marine warm-blooded animals in the North and Baltic Seas: Foundations for assessment of offshore wind farms), were created and accomplished by the University of Kiel's Research and Technology Centre West Coast (RTC) at Büsum, the Leibniz Institute of Marine Sciences at Kiel, the German Oceanographic Museum at Stralsund, the Ruhr University of Bochum (only MINOS), the German Federal Research Centre for Fisheries at Hamburg (only MINOS) and the Administration for the National Park "Wadden Sea of Schleswig-Holstein" at Tönning (Table 1). The scientific competence of the involved institutions for birds and/or mammals of the marine environment was previously established.

The MINOS network identified two main question complexes which should be answered by their research:

- Preferential habitats and migratory routes of animals in the EEZ: Why do they stay at which places during the seasons and where do they move to in between?
- Sense of hearing of porpoises and seals and its sensitivity: What exactly do they hear, how do they react to noise and what kind of noise may damage their ears?

The MINOS research network is comprised of seven sub-projects (SP) to answer these questions for different marine mammals and seabirds.

Table 1: MINOS Research network – different sub-projects

Subproject	**SP1**	**SP2**	**SP3**	**SP4**	**SP5**	**SP6**	**SP7**
Lemma	**Accoustic**	**Abundance**	**Spatial usage**	**Comparison of methods**	**Seabirds**	**Seals at sea**	**Coordination, database**
University of Kiel/RTC Büsum	X	X		X	X	(X)	
German Oceanic Museum			X	X			
Leibniz Institute of Marine Science						X	
National Park Administration				(X)			X
*University of Bochum	X						
*Federal Research Centre for Fisheries		X					
Harbour porpoise	X	X	X	X			X
Harbour seal	X					X	X
Seabirds					X		X
Audiometry	X						
Transsects by		**airplane**		**airplane, ship**	**airplane, ship**		
POD			X	X			
Hydrophone				X			
Telemetry						X	
* MINOS (2002 to 2004) only							

SP1 Investigations on the influence of accoustic emissions of offshore wind energy plants on marine mammals

At water depths below 10m the wind plants will be installed on pile-foundations. The currently preferred method for bringing the piles into the seabed is piledriving. More than 1000 impacts are needed to install one pile. Windmill-related sound emissions raised concern about the potential acoustic impact on marine mammals in German waters. Harbour seals (*Phoca vitulina*) and harbour porpoises (*Phocoena phocoena*) have been identified as main target species in this respect. This sub-project contains an acoustic study to identify the hearing sensitivity of harbour seals and harbour porpoises in captivity as well as free-ranging harbour seals of the Wadden Sea. The audiometry used is quite similar to that in human medicine. Since harbour porpoises have an ultrasonic location system, they are sensitive to under water noise. This study yields information on whether operational sounds of offshore wind turbines and the noise during the construction phase may cause behavioural changes or even physical harm for both mammal species.

SP2 Investigation of density and distribution patterns of harbour porpoises (*Phocoena phocoena*) in German North and Baltic Seas

To assess the planned construction of offshore wind farms, it was necessary to investigate German waters in respect to distribution and density of harbour porpoises. They are counted from a two-engine airplane flying at low altitude on pre-defined tracks. Complex mathematical models lead to estimations of total numbers of animals within the respective area. The results enable identification of areas of ecological importance for harbour porpoises in the German North and Baltic Seas such as hot spots during the seasons or areas of special importance like calving areas.

SP3 Investigations on the spatial use of harbour porpoises in North and Baltic Seas by means of acoustic methods (PODs).

Harbour porpoises have traditionally been monitored visually from airplane or ships. However, there are times when vision is obstructed due to inclement weather or insufficient light, making the detection of harbour porpoises in particular very difficult. Since harbour porpoises use their ultrasonic location system, it is possible to detect them with passive acoustic methods. Within the framework of MINOS, the new acoustic method POD (Porpoise Detector) has been tested and used for investigating habitat use of harbour porpoises in the Baltic Sea. PODs were placed in coastal waters of Mecklenburg-Western Pomerania and Schleswig-Holstein as well as in the German EEZ. PODs deliver data over a period of several weeks and permit analysis at one place over time. The selection of POD stations allows the sampling of information on spatial requirements of harbour porpoises.

SP4 Operation and intercalibration of visual and acoustic methods for assessing occurrence of harbour porpoises (*Phocoena phocoena*)

Visual and acoustic methods are now available for observing harbour porpoises (see subprojects 2 and 3). Visual methods are generally designed to investigate spatial distributions over a short time span. The use of PODs is an acoustic method designed to investigate long-term changes in porpoise abundance in a small area. These fundamental differences make a direct comparison of their results difficult. The whale sanctuary of the National Park „Schleswig-Holsteinisches Wattenmeer“ off the islands Sylt and Amrum is a suitable location for a comparison study since the porpoise density is relatively high (Scheidat & Siebert 2003). By using the different methods at the same time in the same area, their comparability can be tested. Furthermore, the practical use of PODs as a routine tool for monitoring porpoises in the rougher area of the North Sea needs to be determined.

SP5 Marine birds and offshore wind energy: analysis of potential conflicts in the German parts of North Sea and Baltic Sea (MINOS) <u>and</u> Spatio-temporal variability of seabird occurrence in German North and Baltic Seas and their assessment concerning offshore wind energy usage (MINOS+)

Among all research and monitoring activities on resting seabirds in the German seas MINOS focuses on divers, sea ducks and gulls. Some of their main resting habitats in the North and Baltic Seas are located within the German EEZ. Focus of the first two years (2002-2004) of research was on identifying distribution and abundance of marine birds in the study area

(Garthe et al. 2004). Standardised counts on pre-defined transects from ships and - for the first time in Germany - from airplanes were used. Distribution maps and comments are given for 26 seabird species, covering up to four seasons. On this basis, further analysis - including other existing data - allow detection and description of the variability of species occurrence in time and space. Once the ecology of distribution, abundance and variability of seabirds is understood, assessment tools to minimise conflicts between seabirds and human activities can be devised (e.g. Garthe & Hüppop 2004).

SP6 Seals at sea – investigations on spatial and temporal patterns of seals in the North Sea

Telemetric methods are used to fill the gaps in the knowledge on harbour seals mentioned above. Once glued to a seal, the telemetric unit records data of diving and feeding activities for several weeks before it detaches automatically. Three resting sites of seals with different characteristics were chosen. The first is situated on a typical tidal flat area, the second is a real offshore site and the third is situated close to the existing wind farm at Horns Rev, Denmark. The analysis of the data yields information on spatial and seasonal variability as well as ecological patterns during their trips.

SP7 Network coordination and MINOS database

The long-term storage and documentation of all original project data and results are top priorities for the project database located at the National Park Administration. Development of a relational database will provide long-term access to the data of the MINOS network, central availability of analyses, interfaces for data exchange (e.g. with the Seabirds-at-Sea database) and traceability and quality assurance of the results. Data models developed for the MINOS database had to be usable for future monitoring and analysis procedures as illustrated in subproject 4.

Effects of MINOS

MINOS is embedded in an overall process that is currently of social relevance and will remain so in the future. Therefore, the work of the MINOS network was not only geared to answer scientific questions on the core issues (see following chapters) but also to ad hoc demands or ongoing processes in related activities.

There was thus an intensive cooperation with the University of Berlin (project: Instruments for the environmental planning concerning the approval of offshore wind farms), particularly with regard to the selection of criteria (subjects of protection) and specifications (Köller et al. 2006).

The selection and suggestions for designation of Natura 2000 marine protection sites by the German Federal Agency for Nature Conservation (BfN) took place concomitantly to MINOS (von Nordheim et al. 2006). The basis for these suggestions was mainly data from three BfN projects. Inevitable deficits, however, could be compensated by MINOS. On the other hand, data from the BfN projects also served as important additions to the data and findings of MINOS.

MINOS provided essential administrative assistance during the current approval procedures for offshore wind farms in Germany. MINOS was able to provide information on request, using data or the knowledge of experts working in the research network. Furthermore, they contributed to the identification of methods and standards necessary to apply for offshore wind farm building permits (BSH 2007).

Long-term storage of the data is a new challenge for the offshore area. The solutions of the MINOS database should therefore be taken into account to save valuable data and fulfill future reporting obligations (e.g. Natura 2000).

References

Federal Agency for Nature Conservation (BfN)(2004). Natura 2000 sites nominated according to EU habitat directive and the birds directive in the German Exclusive Economic Zone (EEZ). http://www.habitatmare.de/en/downloads/erlaeuterungstexte/Map1_NATURA2000_sites_with_coordinates.pdf

Federal Maritime and Hydrographic Agency (BSH)(2007). Standard Untersuchung der Auswirkungen von Offshore-Windenergieanlagen auf die Meeresumwelt (StUK 3), 58 pp.

Federal Maritime and Hydrographic Agency (BSH)(2007). Continental Shelf Research Information Sytem (CONTIS). http://www.bsh.de/de/Meeresnutzung/Wirtschaft/CONTIS-Informationssystem.

Garthe S & Hüppop O (2004). Scaling possible adverse effects of marine wind farms on seabirds: developing and applying a vulnerability index. Journal of Applied Ecology 41:724-734.

Hammond PS, Benke H, Berggren P, Borchers DL, Buckland ST, Collet A, Heide-Jørgensen MP, Heimlich-Boran S, Hiby AR, Leopold MP, Øien N (1995). Distribution and abundance of the harbour porpoise and other small cetaceans in the North Sea and adjacent waters. Final Report, LIFE 92-2/UK/027, 242 pp.

Kempf N, Todt P, Hälterlein B, Eskildsen K (2000). Trischen – Perle im Nationalpark. Schriftenreihe des Nationalparks Schleswig-Holsteinisches Wattenmeer, Heft 13. Boyens, Heide, 40 pp.

Köller J, Köppel J, Peters W (Eds.)(2006). Offshore wind energy – Research on Environmental impacts. Springer, Berlin, Heidelberg, 371 pp.

von Nordheim H, Boedecker D, Krause JC (Eds.)(2006). Progress in Marine Conservation in Europe – Natura 2000 sites in German Offshore Waters. Springer, Berlin, Heidelberg, 263 pp.

Scheidat M & Siebert U (2003). Aktueller Wissensstand zur Bewertung von anthropogenen Einflüssen auf Schweinswale in der deutschen Nordsee. Seevögel. Zeitschrift Verein Jordsand. Hamburg 24 (3):50-60.

Trilateral Seal Expert Group (2006). Aerial Surveys of harbour and grey seals in the Wadden Sea in 2006. - http://www.waddensea-secretariat.org/news/news/Seals/Annual-reports/seals2006.html.

Garthe S, Dierschke V, Weichler T, Schwemmer P (2004). Teilprojekt 5 - Rastvogelvorkommen und Offshore-Windkraftnutzung: Analyse des Konfliktpotenzials für die deutsche Nord- und Ostsee. In: Kellermann A, Eskildsen K, Frank B (Eds.): Marine Warmblüter in Nord- und Ostsee: Grundlagen zur Bewertung von Windkraftanlagen im Offshore-Bereich, Endbericht zum F&E-Vorhaben (FKZ: 0327520), http://www.minos-info.de/minos1_download_rep.htm.

Excursus 1: Harbour porpoise

Katrin Wollny-Goerke, Ursula Siebert

The **harbour porpoise** (*Phocoena phocoena*) belongs to the suborder *Odontoceti* (toothed whales). It is one of the smallest cetacean species (1.6-1.8m of length, females being larger than males) and inhabits coastal or shelf waters of the northern hemisphere, including the North and Baltic Seas. Genetic and morphological investigations showed that harbour porpoises from the German North and Baltic Seas belong to three different subpopulations: southern and central North Sea, Western Baltic (including Kattegat, inner Danish and German waters) and eastern Baltic.

Harbour porpoises can weigh up to 70kg maximum. They normally reach sexual maturity at the age of 3 to 4 years. After 10 months of gestation they give birth in June / July to a single calf of 65 to 75cm length. The lactation period is approximately 10 months, but newborns begin to feed on fish at approximately 5-6 months of age.

Harbour porpoises have a seasonally fluctuating, but overall very high metabolism. This high energy demand results in part from the fact that the body temperature of these warm-blooded animals also has to be maintained under low water temperature conditions. In addition, chasing single, quick prey fish requires a lot of energy. Moreover, most females are often both pregnant and lactating at the same time, and thus constantly require sufficient and nutritious food. Harbour porpoises are opportunistic feeders. In the North Sea they feed primarily on common sole and sandeels, in the Baltic Sea gobies seems to be the most important prey.
Harbour porpoises are threatened by a variety of anthropogenic impacts such as by-catch, drowning in gillnets, habitat degradation, chemical pollution, food depletion and noise pollution.
The harbour porpoise uses echolocation as an active sensory system for detection of food, obstacles, predator avoidance, navigation, possibly for communication, etc. and has a very sensitive sense of hearing. The animals are therefore highly susceptible to the effects of sound emissions, which can range from mild disturbance to impairment of foraging, hearing loss and even death.
By-catch in fisheries constitutes the largest threat to harbour porpoises in the North and Baltic Seas, where several thousand animals are killed each year.

Particularly the Eastern Baltic population is considered highly endangered. The population size is estimated to be only several hundred individuals. By-catch from fisheries exerts continuous pressure on the population. To address these issues, the "Recovery plan for the Baltic harbour porpoise" (Jastarnia Plan) was adopted by the ASCOBANS (Agreement on the Conservation of Small Cetaceans of the Baltic and North Seas) states to prevent its extinction.

The results of aerial surveys illustrate clear population centres for these small cetaceans. In the EEZ of the North Sea, the region of the Sylt Outer Reef protected area is of high importance, as it is clearly a significant calving and mating habitat in connection with the Harbour porpoise conservation area off Sylt (see chapter 2). The analysis of the acoustic records from detection devices (PODS – POrpoise Detectors) in the Baltic Sea reveals the presence of harbour porpoises in regions that were not so obvious from the aerial or ship counts, for example in the Kadet Channel. There is an observable decline in abundance in the Baltic Sea from west to east (see chapter 2).

2 Harbour porpoises – abundance estimates and seasonal distribution patterns

Anita Gilles, Helena Herr, Kristina Lehnert, Meike Scheidat, Ursula Siebert

Zusammenfassung

Im Rahmen der MINOS und MINOS+ Projekte wurden vom Forschungs- und Technologiezentrum Westküste (FTZ) im Zeitraum Mai 2002 bis Juni 2006 Flugzählungen in der Ausschließlichen Wirtschaftszone (AWZ) und der 12 sm Zone von Nord- und Ostsee durchgeführt. Die Hauptziele dieser beiden Verbundprojekte bestanden in einer Abundanzabschätzung, Untersuchung der Verteilungsmuster sowie einer Analyse der räumlichen und saisonalen Unterschiede in Dichte und Verteilung. Mit dem Flugzeug, einer Partenavia 68 ausgerüstet mit sog. „Bubble"-Fenstern, wurde auf einer Fläche von 64.000 km^2 eine Strecke von insg. 62.000 km auf Transektlinien zurückgelegt. Insgesamt wurden 3.099 Schweinswalgruppen in den Untersuchungsgebieten der Nordsee und 230 in der Ostsee gesichtet. Die geschätzte Abundanz in der Nordsee war am höchsten im Mai/Juni 2006 mit geschätzten 51.551 Tieren (%VK=32) und im April/Mai 2005 mit 38.089 Tieren (%VK=38). Demnach ist im Spätfrühling und Frühsommer mit den höchsten Dichten zu rechnen. Geringste Zahlen wurden für den Herbst berechnet, z. B. 11.573 Tiere (%VK=34) im Oktober/November 2005. Bezüglich der räumlichen Verteilung zeigen die Ergebnisse der MINOS-Erfassungen sehr deutlich, dass die Habitatnutzung heterogen ist, wobei Schweinswale klare Präferenzen für einige bestimmte Gebiete zeigen, die daher als wichtige Gebiete zur Nahrungssuche angesehen werden können. Diese Präferenz ist am deutlichsten im Frühling und Sommer, als Hot Spots im Bereich des Borkum Riffgrundes (nur im Frühling) und des Sylter Außenriffs (Frühling und Sommer) entdeckt wurden, und weniger ausgeprägt im Herbst. Im Gegensatz zur Nordsee, wurden die Dichten in der Ostsee ungefähr 10mal geringer geschätzt. Die Abundanz wurde am höchsten geschätzt im Juni 2005 (2.905; %VK=41) und im September 2005 (2.763, %VK=41). Da die Abundanzschätzung für die Erfassung im September 2004 auch in über 2.500 Tieren resultierte, wird angenommen, dass die Dichte im Spätsommer/Frühherbst am höchsten ist. Die Schweinswaldichte weist im Trend einen West-Ost-Gradienten auf, mit höchsten Dichten in der Kieler und Mecklenburger Bucht und sehr geringen Dichten in der Pommerschen Bucht.

Abstract

Within the scope of the MINOS and MINOS+ projects, the Research and Technology Centre Westcoast (FTZ) conducted aerial line transect sighting surveys that covered waters in the German Exclusive Economic Zone (EEZ) and the 12 nautical mile zone of the North and the Baltic Sea from May 2002 to June 2006. The main aims of MINOS and MINOS+ were to estimate abundance of harbour porpoises, investigate their distribution patterns and analyse spatial and seasonal differences in density and distribution. The airplane, a Partenavia 68 equipped with bubble windows, surveyed 62,000km of track lines within an area of 64,000km^2. A total of 3,099 harbour porpoise sightings was recorded on effort in the North and 230 in the Baltic Sea strata. The estimated abundance in the **North Sea** was highest in May/June 2006 with an estimate of 51,551 animals (%cv=32) and in April/May 2005 with an estimate of 38,089 animals (%cv=38). Thus, density was highest in late spring to early summer. Lowest numbers were estimated in autumn, e.g. 11,573 animals (%cv=34) in Oct./Nov. 2005. In terms of spatial distribution, the results of the MINOS surveys show very clearly that area use of harbour porpoises is heterogeneous, with the animals showing clear preferences for several discrete areas, suggesting that these may be important foraging grounds. The preference is most clear in spring and summer, where hot spots were detected in the area of *Borkum Reef Ground* (spring only) and *Sylt Outer Reef* (spring and summer), and less evident in autumn. In contrast to the North Sea, densities in the **Baltic Sea** study area were about 10 times lower. Abundance was estimated to be highest in June 2005 (2,905; %cv=41) and September 2005 (2,763; %cv=41). Lowest numbers were estimated in March/April 2005 (1,352; %cv=61) and April 2006 (1,635; %cv=45). As the survey in September 2004 also resulted in an abundance estimate of more than 2,500 animals, it is concluded that densities are highest in late summer/early autumn (both September surveys were conducted in the first week of the month). The density of porpoises shows a declining trend from west to east, with highest densities in the areas Kiel and Mecklenburg Bight and very low densities in Pommeranian Bay.

Introduction

The harbour porpoise (*Phocoena phocoena*) is the smallest cetacean inhabiting temperate to cold waters throughout the northern hemisphere. In the North Atlantic, the species inhabits coastal waters of Europe, of North America and of northern Africa. The harbour porpoise is the only cetacean species found regularly in both the German North and Baltic Seas (Reijnders 1992, Benke et al. 1998, Hammond et al. 2002, Scheidat et al. 2004, Siebert et al. 2006). Due to its occurrence, mainly but not exclusively, in coastal or shelf waters the porpoise is threatened by a variety of anthropogenic impacts (Hutchinson et al. 1995, Kaschner 2001, Scheidat & Siebert 2003), including by-catch in fishery (Kock & Benke 1996, Vinther 1999, Lockyer & Kinze 2003, Vinther & Larsen 2004) and habitat degradation due to e.g. chemical pollution (Jepson et al. 1999, Siebert et al. 1999).

To evaluate further anthropogenic impacts and their cumulative effects, e.g. the planned construction of offshore windmill farms, it is necessary to investigate German waters with respect to recent distribution and density of harbour porpoises. Until recently very little data existed on distribution and abundance in the German North and Baltic Seas. Most information was based on results of the SCANS survey from July 1994 (Hammond et al. 2002). In 1994, harbour porpoise abundance was estimated to be 341,366 animals (%cv=14; 95% CI = 260,000-449,000) in an area of 1mio. km^2. Eleven years later, in July 2005, SCANS II took place. No difference was found in the abundance of harbour porpoises in 1994 and 2005. However, in 2005 the average density in survey blocks north of 56°N was approximately half the density estimated in 1994, and the average density in survey blocks south of 56°N in 2005 was approximately twice the density estimated in 1994. Both these differences are significant at the 5% probability level (P. Hammond, pers. comm.).

SCANS and SCANS II were both large-scale surveys that aimed at a synoptic coverage of a large survey area. No small-scale surveys were conducted in German waters that covered the complete EEZ and the 12nm zone over the course of many years.

Thus, the aims of MINOS and MINOS+ were to:

- estimate abundance of harbour porpoises in the German EEZ and 12nm zone;
- investigate distribution patterns;
- analyse spatial and seasonal differences in density and distribution.

Methods

Study area and aerial survey methodology

Intensive aerial surveys were conducted all year round between May 2002 and June 2006 to assess distribution and abundance of harbour porpoises in the German Exclusive Economic Zone (EEZ) and the 12 nautical mile zone of the North and the Baltic Sea. The study area in the south-western Baltic Sea was extended to the Danish Isles. Due to logistical constraints (e.g. plane refuelling) the study area was divided into seven strata prior to conducting the surveys (Fig. 1).

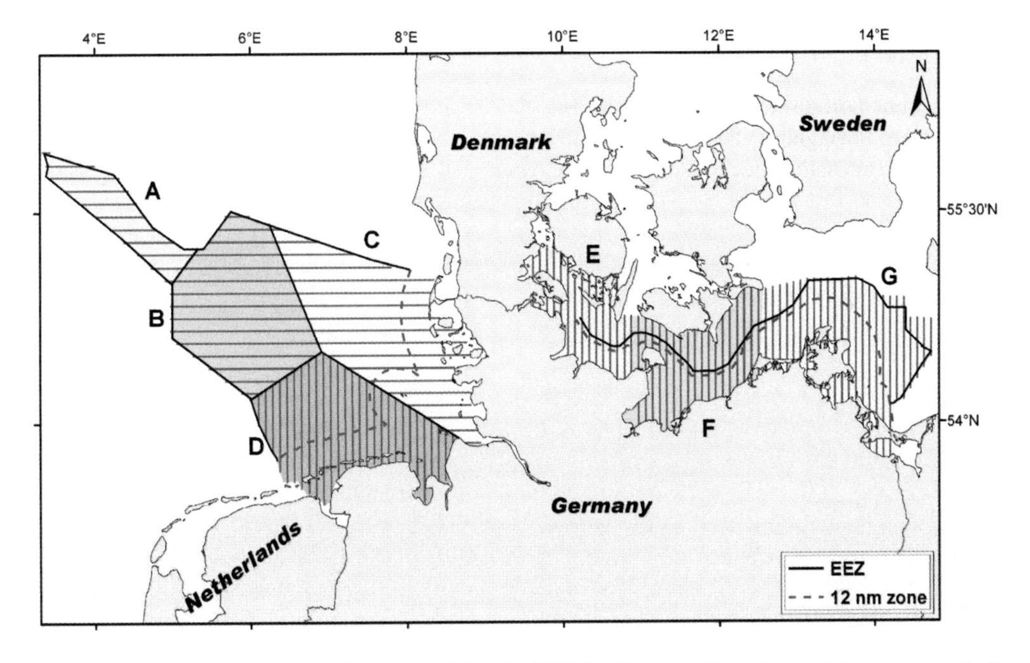

Figure 1: Study areas of the projects MINOS and MINOS+. Transect lines for aerial surveys are indicated by the solid lines. Transect lines are equispaced: 10 km in the North Sea (except area D with 6 km space) and 6 km in all Baltic Sea strata.

The size of the total study area was 41,000km^2 in the North Sea and 23,000km^2 in the Baltic Sea.

The methodology followed standard line transect distance sampling techniques, where the observer travels along a line, recording detected objects and the distance from the line to each object (Hiby & Hammond 1989, Buckland et al. 2001). Line transect sampling is one of the most widely used techniques for estimating the size of wildlife populations. The survey design is comprised of a grid of systemically spaced transect lines randomly superimposed on the study area. As transect direction should not parallel some physical or biological feature to avoid an unrepresentative sample, transects were placed either in east-west or north-south direction (Fig. 1) to run perpendicular to water depth gradients, as recommended by Buckland et al. (2001).

Surveys were flown at 100 knots (185km/h) at an altitude of 600 feet (183m) above the water surface in a Partenavia P68, a two-engine, high-wing airplane, equipped with two bubble windows to allow scanning directly underneath the plane. The survey team consisted of two observers, one data recorder (navigator) and the pilot. Sighting data were acquired by the two observers at the time, each positioned on one side of the airplane at a bubble window, scanning for animals with the naked eye. The navigator entered all reported data online into a toughbook computer interfaced with a global positioning system (GPS). The computer stored the airplane's position every 2s. Additionally, the start and end positions of the transect lines and the sighting positions were recorded.

Surveys were only conducted when Beaufort sea state conditions were less than or equal to three and when visibilities were greater than 3km. Environmental conditions were recorded at the beginning of each transect and were updated with any change. Data collected on environmental conditions included Beaufort Sea state, water turbidity, percentage of cloud cover and for each observation position, glare (magnitude of glare and angle obscured by glare) and overall viewing quality (good, moderate or poor). All data recorded in poor sighting conditions (e.g. Beaufort Sea state > 3) were excluded from subsequent analysis.

Data recorded for each harbour porpoise sighting included:

- exact position when the group passed perpendicular to the window
- angle of inclination to the group, measured by a hand-held inclinometer
- estimated group size
- number of calves
- sighting cue
- behaviour
- swim direction

The perpendicular distances from the transect to the group were later calculated from airplane altitude and declination angle.

During the census of cetaceans it is never possible to detect all objects present on the transect line. Two factors influence detection: i) availability bias and ii) perception bias (Marsh & Sinclair 1989, Laake et al. 1997). These two factors are usually combined in g(0), which is the probability of detecting an object on the transect line (Buckland et al. 2001). The availability bias describes the probability of an animal being available for detection (i.e., at or near the surface) whereas the perception bias incorporates the probability that an animal, when it is physical available for detection, is missed by the observer (e.g. due to fatigue, observer experience or environmental condition). For determining g(0) the Hiby racetrack data collection method was used, which involves some doubling-back to re-survey previously flown transect segments (Hiby & Lovell 1998, Hiby 1999). From the time and position of the original as well as the re-sightings, the Hiby algorithm determines the probability that the sightings are the same group. These probabilities are then used to estimate g(0).

The effective half-strip width (esw) is the width of the area searched effectively on each side of the line transect (Buckland et al. 2001). Esw was calculated separately for good and moderate conditions and separately for right and left observer sides.

Abundance estimates

The density of harbour porpoises was calculated using line transect estimates based on perpendicular distances of sightings to the transect.

Animal abundance ($\hat{N}_v$) in stratum v was estimated as:

$$\hat{N}_v = \frac{A_v}{L_v}\left(\frac{n_{gsv}}{\hat{\mu}_g} + \frac{n_{msv}}{\hat{\mu}_m}\right)\overline{s}_v$$

Where A_v is the area of the stratum, L_v is the length of transect line covered on-effort in good or moderate conditions, n_{gsv} is the number of sightings that occurred in good conditions in the stratum, n_{msv} is the number of sightings that occurred in moderate conditions in the stratum, $\hat{\mu}_g$ is the estimated total effective strip width in good conditions, $\hat{\mu}_m$ is the estimated total effective strip width in moderate conditions and $\overline{s}_v$ is the mean observed school size in the stratum.

Densities were estimated by dividing the abundance estimates by the area of the associated stratum. The coefficient of variation (CV) and the 95% confidence intervals (CI) were estimated by bootstrapping (999 replicates) within strata, using transects as the sampling units.

Distribution mapping

Data of the six study years (2002-2006) were pooled across seasons since prior statistical tests (Pearson's correlation) did not detect any significant variation between data collected in different years. Seasons were defined according to meteorological divisions (spring: March to May, summer: June to August, autumn: September to November). The winter months were excluded due to poor effort and thus insufficient coverage of the study area.

For the spatial analysis in ArcGIS 8.3 (ESRI) a grid with a resolution of 10x10km was created, dividing the study area into 550 quadrate cells (e.g. North Sea). The line transect point locations were converted into mean density estimates per grid cell ($\hat{D}$ [indiv. km^{-2}]):

$$\hat{D} = \frac{n_{indiv.}}{effort}$$

where $n_{indiv.}$ is the sum of harbour porpoises per cell and $effort$ is the area searched effectively (in km^2).

Results

In the study period of six years 62,000km of track lines were surveyed within a region of 64,000km^2: Of these, 37,000km were flown in the North Sea and 25,000km in the Baltic Sea. A total of 3,099 harbour porpoise sightings were recorded on effort in the North Sea and 230 in the Baltic Sea strata. Data were used to calculate abundance estimates (part I) and to produce maps of harbour porpoise seasonal distribution patterns (part II).

Abundance estimates

MINOS and MINOS+ aimed at covering all seven strata four times a year in order to account for seasonal variability. A survey of the total study area, in either North or Baltic Sea, was considered to be complete, when a representative coverage of the transect lines in good or moderate conditions was achieved in a short period of time. It was then possible to calculate abundance estimates. Complete surveys of the North Sea strata were achieved in 41 consecutive days' maximum and in the Baltic Sea in 25 consecutive days. In Tables 1 and 2 only surveys with a CV lower than 0.7 are presented as these were considered to be robust estimates.

Due to unfavourable weather conditions it was not always possible to achieve a good coverage of stratum A. Thus, for better comparison between surveys, the six abundance estimates in Table 1 refer to data collected in strata B, C and D only. The abundance estimates ranged between 11,573 animals in Oct./Nov. 2005 (95% CI = 6,077-22,222) and 51,551 animals in May/June 2006 (95% CI = 27.879-98,910).

Table 1: Estimates of harbour porpoise abundance for the North Sea study area (strata B, C & D). CV=coefficient of variation, CI=95% confidence interval

survey	effort (km)	abundance	CV	CI low	CI high
Sep./Oct. 2002	1,995	22,562	0.44	9,112	49,850
March/April 2003	1,296	17,556	0.37	8,791	36,231
April/May 2005	2,476	38,089	0.38	19,628	81,126
Aug./Sep. 2005	2,352	17,618	0.38	8,786	36,574
Oct./Nov. 2005	3,108	11,573	0.34	6,077	22,222
May/June 2006	2,696	51,551	0.32	27,879	98,910

In the Baltic Sea study area, six surveys met the criteria mentioned above. The resulting abundance estimates are listed in Table 2. The estimates ranged between 1,352 animals in March/April 2005 (95% CI = 230-3,840) and 2,905 animals in June 2005 (95% CI = 1,308-6,384).

Table 2: Estimates of harbour porpoise abundance for the Baltic Sea study area (strata E, F & G). CV=coefficient of variation, CI=95% confidence interval

survey	effort (km)	abundance	CV	CI low	CI high
June 2003	1,921	1,726	0.39	778	3,750
Sep. 2004	2,693	2,547	0.36	1,312	5,461
March/April 2005	1,922	1,352	0.61	230	3,840
June 2005	3,244	2,905	0.41	1,308	6,384
Sep. 2005	2,337	2,763	0.41	1,193	5,902
April 2006	2,492	1,635	0.45	607	3,560

Seasonal distribution of harbour porpoises

The main aerial survey results per season are shown in Table 3. As the effort between seasons was similar in North and Baltic Seas, it is possible to compare distribution patterns.

Table 3: Effort summary per season and main aerial survey results in the North (NS) and Baltic Sea (BS) strata. Each season was pooled over the years 2002 to 2006.

	Flight days	Track line length (km)	Effort (km²)	No. of groups	No. of individuals	No. of calves
spring_NS	23	10,905	1,178	1,448	1,689	19
summer_NS	27	12,847	1,179	1,256	1,605	121
autumn_NS	25	13,388	1,172	395	520	28
total_NS	**75**	**37,140**	**3,529**	**3,099**	**3,814**	**168**
spring_BS	18	9,130	908	47	72	0
summer_BS	12	8,658	914	125	201	7
autumn_BS	13	7,520	811	58	82	6
total_BS	**43**	**25,308**	**2,633**	**230**	**355**	**13**

Spring North Sea

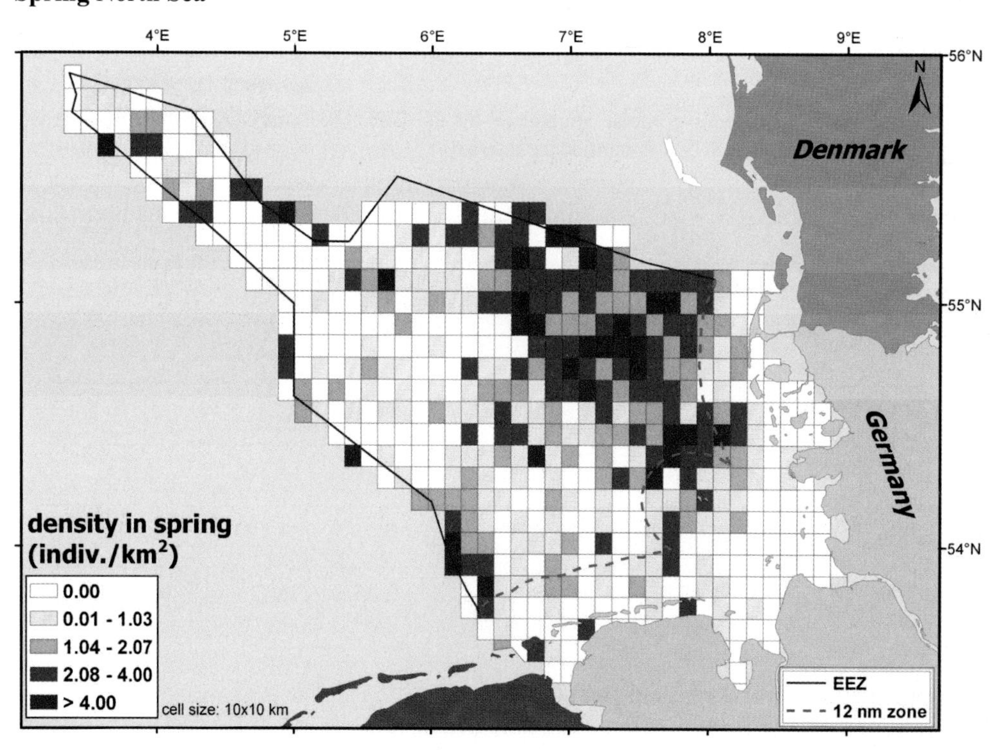

Figure 2: Density distribution of harbour porpoises in spring (March, April, May) in the study area of the North Sea. Data from the study years 2002-2006 were pooled.

In spring, harbour porpoise distribution was highly clumped. The density distribution map shows two "hot spot" areas of very high harbour porpoise densities (Fig. 2). One was located in the south-western part of the German EEZ, approx. 60km offshore of the East Frisian Islands, in an area called *Borkum Reef Ground*. In contrast to the second area of highest harbour porpoise density, this aggregation was comparatively small in size. Offshore the North Frisian Islands of Sylt and Amrum, in an area spanning from 40 to about 130km west of the island of Sylt, harbour porpoise densities were highest. Across all study years this very large aggregation zone was consistently found in the north-eastern part of the German EEZ in an area called *Sylt Outer Reef*. In addition, high densities were also observed in the vicinity of the island of Helgoland and on the Doggerbank in stratum A. The most westerly part of the German EEZ is part of the submerged sandbank Doggerbank. So far no small-scale studies concerning the distribution and abundance of marine mammals have been conducted in that area and the results show that porpoises occur regularly.

Summer North Sea

In summer, a pronounced north-south density gradient was observed (Fig. 3). Again, large aggregations of harbour porpoises were detected in the north-eastern part of the German EEZ.

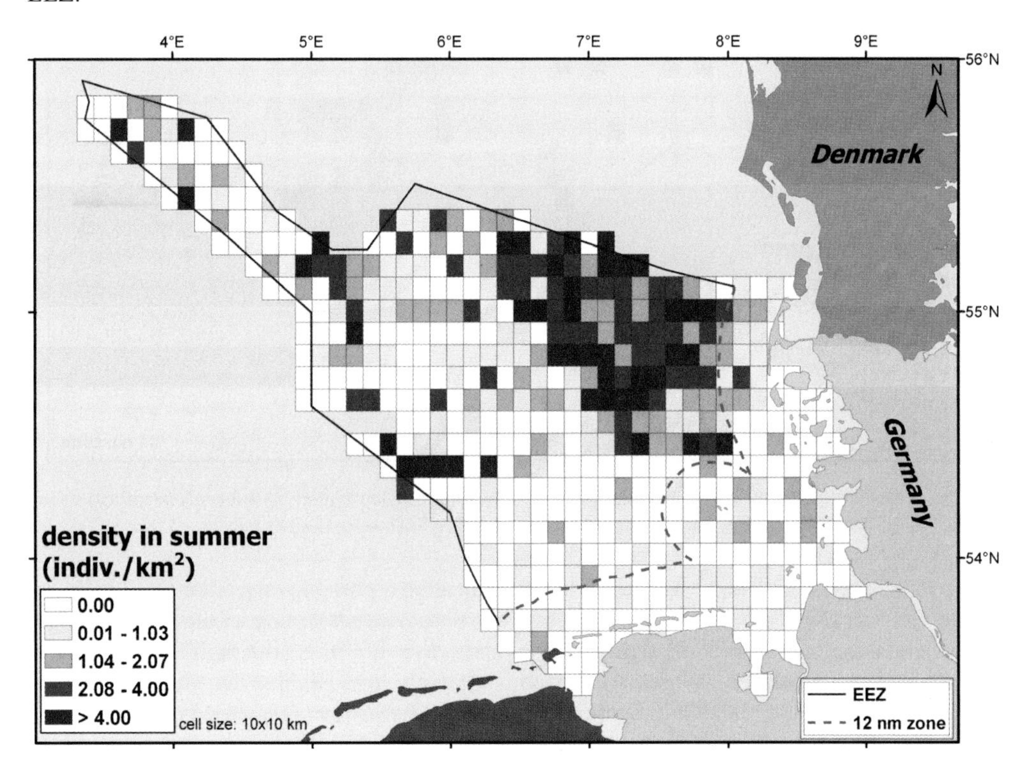

Figure 3: Density distribution of harbour porpoises in summer (June, July, August) in the study area of the North Sea. Data from the study years 2002-2006 were pooled.

In comparison to the spring pattern a shift in the distribution was observed, as significant lower densities were calculated south of 54°30'N, where densities were six times lower than those found in the north during this season.

Autumn North Sea

In autumn, harbour porpoises appeared to be more evenly dispersed throughout the study area (Fig. 4). No specific aggregation area was detected. The sighting rate was lowest in comparison to the other seasons.

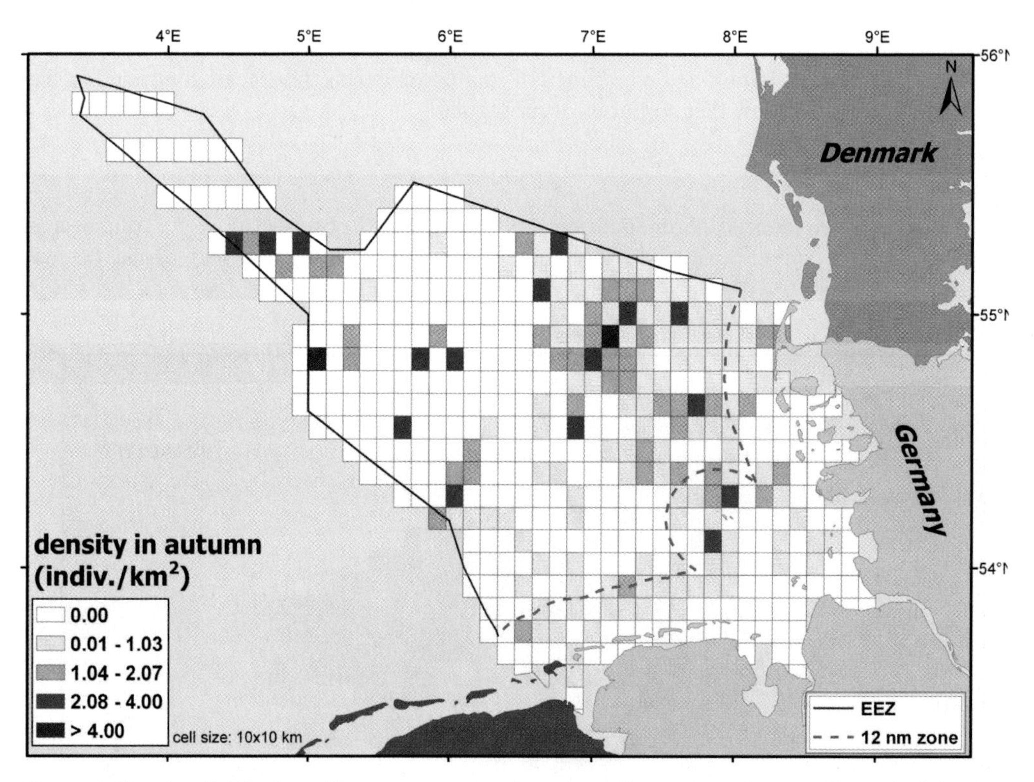

Figure 4: Density distribution of harbour porpoises in autumn (September, October, November) in the study area of the North Sea. Data from the study areas 2002-2005 were pooled.

The overall estimated density was much lower, indicating a possible migration of animals out of the German Bight during autumn. The harbour porpoises remaining in the German Bight were sighted mainly (92% of all sightings) in waters east of 6°30'E, although the coverage in the offshore strata (namely in area A) was comparatively low. Highest densities were found near the island of Helgoland and, again, in the area of *Sylt Outer Reef*. In comparison to the summer months a higher number of porpoises was detected in the waters around the East Frisian Islands, although the density was much lower than in spring.

Spring Baltic Sea

The density of porpoises was highest in the western Baltic Sea, namely in the Kiel Bight and Flensburg Fjord, moreover, around the island of Fehmarn and in the eastern part of the study area close to the Polish border (Fig. 5). However, it should be noted that all sightings east of the island of Rügen occurred only in May 2002 and April 2004. In all other years, despite very high effort in that area, no sightings were recorded in spring. Thus, an enormous change in the use of this area between the years must be assumed. Sighting rates were lowest in survey stratum F.

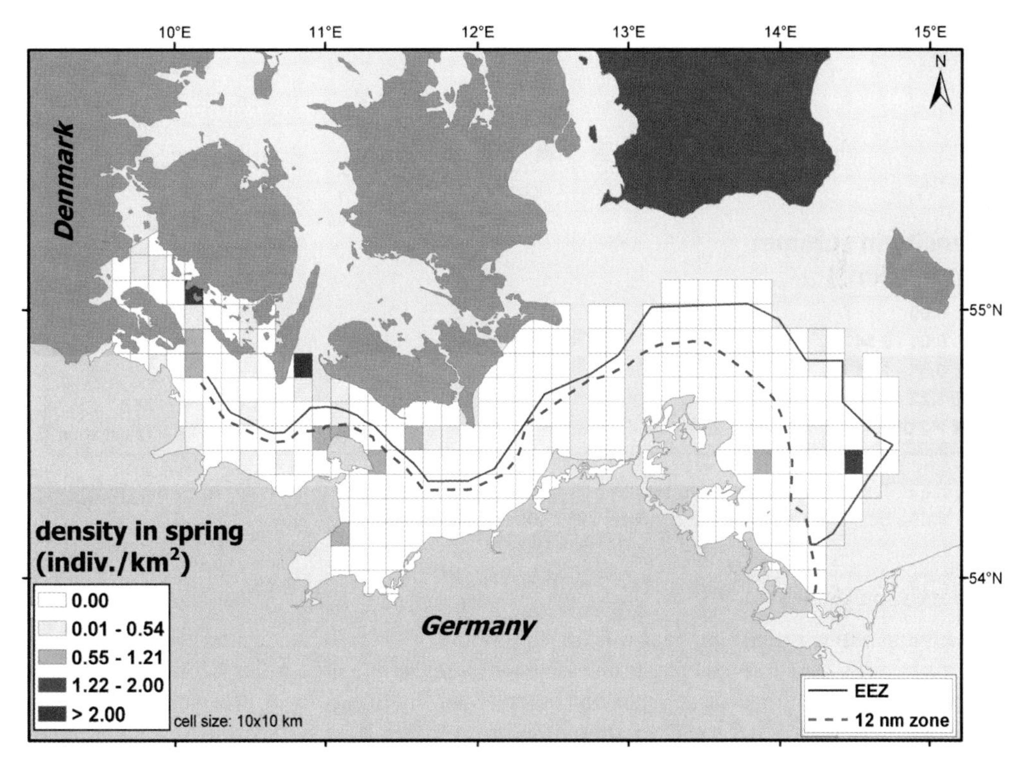

Figure 5: Density distribution of harbour porpoises in spring (March, April, May) in the study area of the Baltic Sea. Data from the study years 2002-2006 were pooled.

Summer Baltic Sea

The highest number of porpoise sightings was achieved during the summer months (Fig. 6). Highest densities were again estimated for stratum E (Kiel Bight), mainly near the Danish Isles. The summer densities in stratum F were higher than in spring. A trend for declining densities from west to east could be observed. Lowest densities were always estimated for stratum G, with the exception of one event in July 2002: an aggregation of 84 porpoises was sighted on the Oderbank in the Pomeranian Bay. In the following three summers no single sighting was recorded in that area.

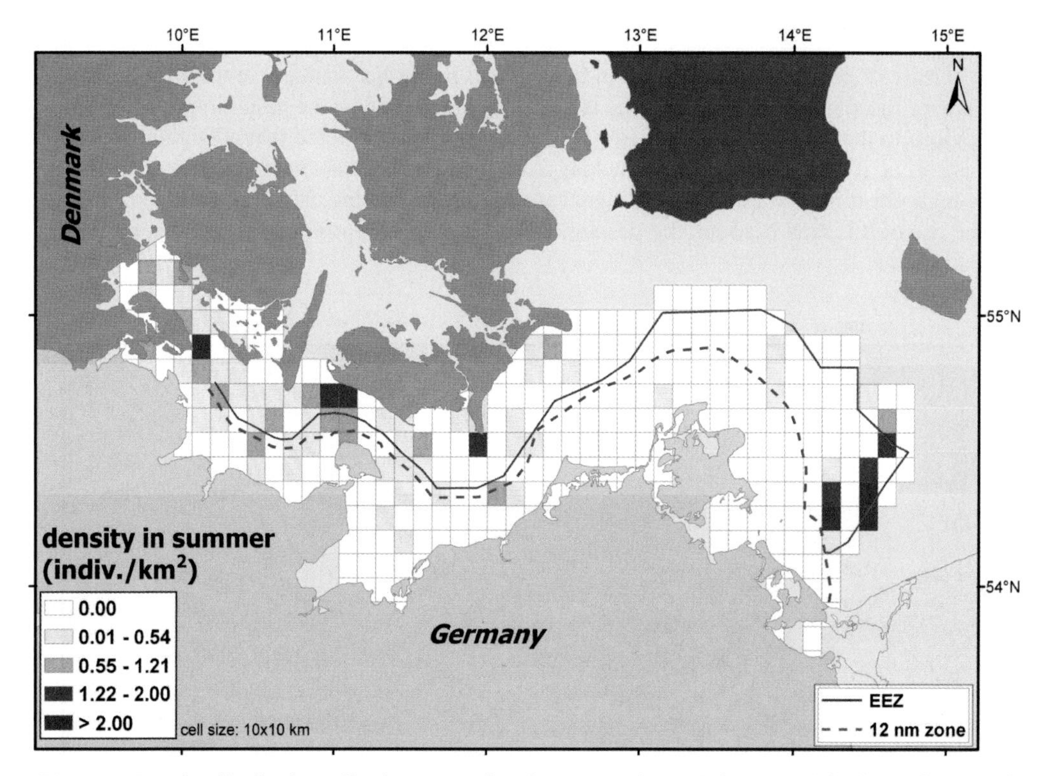

Figure 6: Density distribution of harbour porpoises in summer (June, July, August) in the study area of the Baltic Sea. Data from the study years 2002-2005 were pooled.

Autumn Baltic Sea

In autumn, the west-east gradient was most pronounced (Fig. 7). No sightings were made east of the peninsula Darss, except for one single sighting in the area Adler Ground. The overall density was lower than in summer but higher than in spring. In comparison to spring and summer the north-eastern part of stratum F, close to the border to Sweden, showed higher densities in autumn.

Discussion

MINOS and MINOS+ provided the first abundance estimate for the entire German EEZ and 12nm zone. It was possible to conduct successful surveys in the course of five study years, which could be used to calculate robust seasonal abundance estimates. Most cetacean surveys are only conducted in summer, thus giving a snapshot of a precise time interval only. The data collected in MINOS and MINOS+ provide an excellent opportunity to analyse inter-annual changes as well as seasonal changes in harbour porpoise density and distribution. The results are the baseline for all upcoming surveys in the EEZ and can be used to evaluate anthropogenic impacts in the German North and Baltic Seas.

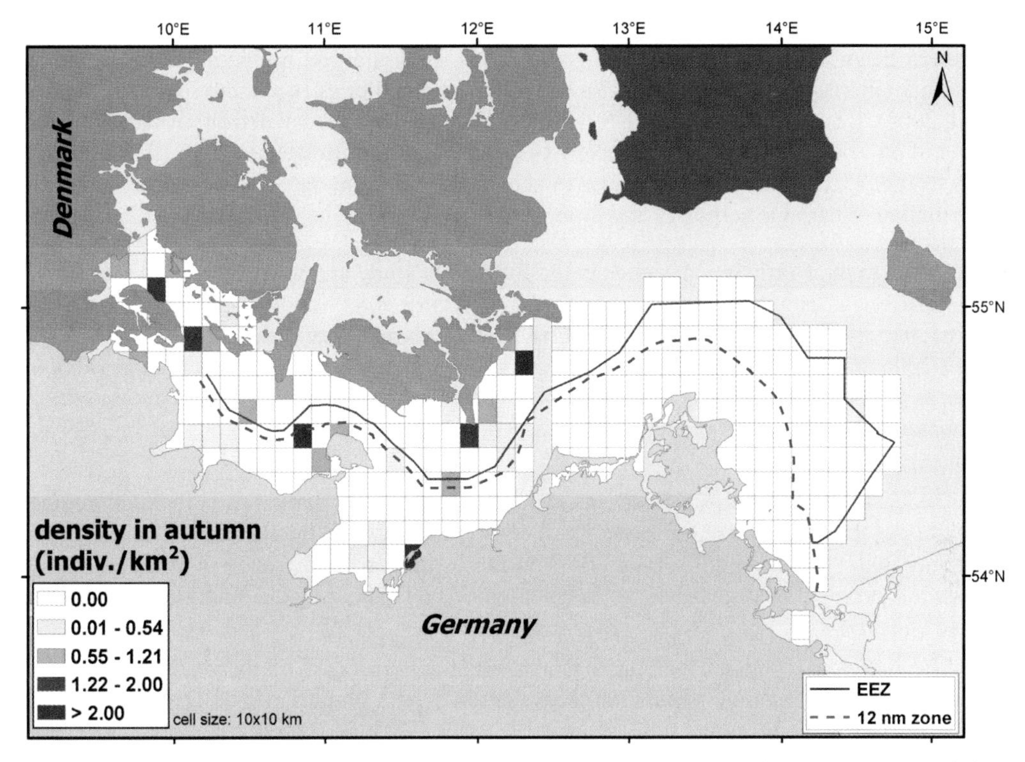

Figure 7: Density distribution in autumn (September, October, November) in the study area of the Baltic Sea. Data from the study years 2002-2005 were pooled.

The abundance in the **North Sea** was highest in May/June 2006 with an estimate of 51,551 animals (%cv=32) and in April/May 2005 with an estimate of 38,089 animals (%cv=38). Thus, highest numbers were found in late spring to early summer. Lowest numbers were estimated in autumn, e.g. 11,573 animals (%cv=34) in Oct./Nov. 2005. This leads to the hypothesis that harbour porpoises migrate into German waters in early spring, reach maximum numbers in early summer and start migrating out of the study area in autumn. It was possible to conduct a few surveys in the course of the winter months, but a complete coverage of the total study area was never achieved. Thus, no abundance could be estimated for winter. However, the surveys in winter were the ones with the lowest sighting rates, completing the picture of the annual migration cycle. The wintering grounds of harbour porpoises remain unknown. Seasonal changes in sex ratio suggest sexual segregation, especially in winter (Lockyer & Kinze 2003).

In terms of spatial distribution, the results of the MINOS surveys show clearly that area use is heterogeneous. Harbour porpoises show clear preferences for several discrete areas, suggesting that these may be important foraging grounds. The preference is most clear in spring and summer and less evident in autumn. In spring, during the onset of migration into German waters, the two hotspot areas *Borkum Reef Ground* and *Sylt Outer Reef* seem to play an important role as key foraging areas from where porpoises spread out. The density of porpoises offshore the East Frisian Islands increased in the course of the study period. From 2004 onwards, mainly in spring, very high sighting numbers were recorded. At the same time the southern neighbouring countries, The Netherlands, Belgium and northern France, were reporting an increase in harbour porpoise strandings and sightings (Camphuysen 2004, Kiszka et al. 2004). Besides that, the SCANS II survey showed that average density in survey blocks south of 56ºN in 2005 was approximately twice the density estimated in 1994 (P. Hammond, pers. comm.). The inter-annual changes in abundance could be due to small-scale changes in distribution of harbour porpoises and/or their prey, or, most likely, some combination of both.

In contrast to the North Sea, densities in the **Baltic Sea** study area were about 10 times lower. Abundance was estimated to be highest in June 2005 (2,905; %cv=41) and September 2005 (2,763; %cv=41). Lowest numbers were estimated in March/April 2005 (1,352; %cv=61) and April 2006 (1,635; %cv=45). Due to overlapping confidence intervals, the situation is not as clear as in the North Sea. As the survey in September 2004 also resulted in an abundance estimate of more than 2,500 animals, it was concluded that densities are highest in early autumn.

Another factor that prevents a clear conclusion concerning the Baltic Sea is the high number of sightings that occurred in the eastern part of the study area, in the Pomeranian Bay, in May and July 2002. Despite high survey effort no more sightings were made in that area during following surveys. Two possible scenarios might explain these single sighting events in the Pomeranian Bay:

(1) The individuals sighted belonged to the subpopulation east of the underwater Darss-Limhamn ridge, a separate population from the rest of the Baltic/Belt Sea (Tiedemann et al. 1996, Börjesson & Berggren 1997, Huggenberger et al. 2002). The original population size of the *Baltic Proper* subpopulation as well as the distribution range of these porpoises have been greatly reduced (Andersen 1982, Skóra et al. 1988, c.f. Koschinski 2002). Thus, the possibility of sighting porpoises of this subpopulation is very low. An aerial survey in July 1995 estimated an abundance of 599 animals (%cv=0.57) for an area around the island of Bornholm (Hiby & Lovell 1996). Possible causes for a decreasing population size include commercial hunting of porpoises (e.g. in Poland and Denmark until the beginning of the World War II; Kinze 1995), high numbers of by-catch in bottom-set gillnets and salmon driftnets (Skóra et al. 1988, Skóra 1991, Berggren 1994, Kock & Benke 1996), increased mortality during severe ice winters (cf. Teilmann & Lowry 1996), habitat degradation (Skóra et al. 1988) as well as impaired health status (Siebert et al. 2001, 2006, Wünschmann et al. 2001), possibly caused by high concentrations of contaminants (e.g. Siebert et al. 1999, Beineke et al. 2005, Das et al. 2006).

Joint activities of ASCOBANS and the IWC have underlined the precarious situation of this stock (e.g. implementation of the recovery plan for Baltic harbour porpoises, Jastarnia plan; ASCOBANS 2002). As part of the Jastarnia project the current status of subpopulations of harbour porpoises inhabiting the Skagerrak, Kattegat, Øresund, Belt Seas and inner Baltic waters were investigated using methods of population genetics. Significant geographical differences were found in the genetic population structure and confirm, as proposed earlier, that porpoises from the eastern inner Baltic (in ICES region IIId, German/Polish/Lithuanian coast) constitute one stock of a separate subpopulation (Tiedemann et al. 2006).

(2) More likely, porpoises from the Belt Sea, which are part of the subpopulation "western Baltic Sea" (incl. Kattegat, Belt Seas, Øresund, Kiel Bight and Fehmarn Belt), followed prey into the Pomeranian Bay. Swarms of herring or sprat may have been detected and followed into waters of the central Baltic. It is known that porpoises although occurring mostly single or in groups of two animals, may temporarily aggregate in large groups when detecting an abundant food supply (Evans 1990). If prey, rich in energy and thus valuable, is only available for a short period of time, such as spawning shoals of herring or sprat, these aggregations of harbour porpoises may be difficult to encounter using standard line-transect methodology in such a low density area as the Baltic Proper.

Apart from this unusual event in July 2002, the density of porpoises declined from west to east during all other study months and years, with highest densities in the areas Kiel and Mecklenburg Bight and lowest densities in Pommeranian Bay. These findings are consistent with the results of the SCANS survey in 1994 (Hammond et al. 2002). In 2005, during SCANSII, no sightings occurred east of the island of Fehmarn (P. Hammond, pers. comm.). The waters east of 19°E were not included in the SCANS study area. However, a Polish ship contributed to the survey effort of project partners from the University of Gdansk and surveyed the Polish Baltic Sea to collect data on relative abundance of harbour porpoises during two weeks in July 2005. In total, the acoustic and visual survey effort amounted to 1,602 km. Although no sightings were recorded, two probable acoustic detections were made. Both detections were isolated clicks characteristic of harbour porpoises.

A decrease in frequency of strandings and incidental sightings along the German coast from Schleswig-Holstein to Mecklenburg-Prepommerania was reported (Siebert et al. 2006). The frequency of porpoise click detections on stationary hydrophones (T-PODs) decreased from west to east as well (Verfuß et al. 2007; also see chapter 3).

The MINOS surveys are an invaluable start to improving our knowledge on harbour porpoise occurrence and density. Continued effort, however, is of the essence. Abundance must be monitored in the future to detect changes in numbers early. As the building of the first offshore windmill farm in German waters is fast approaching, it will be necessary to monitor animal distribution closely in order to follow possible shifts in distribution. As important foraging areas with the highest harbour porpoise densities found in Europe have been identified in our waters, Germany shares the responsibility for the conservation of this species.

Acknowledgements

We would like to thank L. Hiby, P. Lovell and K. Kaschner for their help with data analysis. Our thanks go to P. Siemiatkowski from Syltair, L. Petersen from the Danish Air Survey and S. Hecke from FLM Kiel. Completion of the surveys would have been impossible without the enthusiastic effort of the observers and navigators J. Adams, P. Börjesson, D. Risch, A. Gomez, I. Kuklik, L. Lehnert, M. Marahrens, C. Rocholl, T. Walter, U. Westerberg and S. Zankl. We would like to thank H. Giewat for managing the database and R. Mundry and S. Adler for help with statistical analysis. The MINOS and MINOS+ projects were funded by the German Federal Ministry for the Environment, Nature Conservation and Nuclear Safety (BMU). Data of the project EMSON, funded by the Federal Agency for Nature Conservation (BfN), were used in spatial analysis of distribution patterns.

References

Andersen SH (1982). Change in occurrence of the harbour porpoise, *Phocoena phocoena*, in Danish waters as illustrated by catch statistics from 1834 to 1970. FAO Fish Ser 5, Mammals in the Seas 4: 131-133

ASCOBANS (2002). Agreement on the Conservation of Small Cetaceans of the Baltic and North Seas. Recovery Plan for Baltic harbour porpoises (Jastarnia Plan). Bonn, July 2002.

Beineke A, Siebert U, McLachlan M, Bruhn R, Thron K, Failing K, Müller G, Baumgärtner W (2005). Investigations of the potential influence of environmental contaminants on the thymus and spleen of harbour porpoises *(Phocoena phocoena)*. Environ Sci Technol 39:3933-3938

Benke H, Siebert U, Lick R, Bandomir B, Weiss R (1998). The current status of harbour porpoises (*Phocoena phocoena*) in German waters. Arch Fish Mar Res 46:97-123

Berggren P (1994). Bycatches of the harbour porpoise (*Phocoena phocoena*) in the Swedish Skagerrak, Kattegat and Baltic Seas; 1973-1993. Special Issue 15: Gillnets and Cetaceans. Rep.Int.Whal.Comm. special issue 15:211-215

Børjesson P & Berggren P (1997). Morphometric comparisons of skulls of harbour porpoises (*Phocoena phocoena*) from the Baltic, Kattegat and Skagerrak Seas. Can J Zool 75:280-287

Buckland ST, Anderson DR, Burnham KP, Laake JL, Borchers DL, Thomas L (2001). Introduction to distance sampling. Estimating abundance of biological populations. Oxford University Press, New York

Camphuysen CJ (2004). The return of the harbour porpoise (*Phocoena phocoena*) in Dutch coastal waters. LUTRA 47:113-122

Das K, Vossen A, Tolley K, Vikingsson GA, Thron K, Müller G, Baumgärtner W, Siebert U (2006). Interfollicular fibrosis in the thyroid of the harbour porpoise: an endocrine disruption? Arch Environ Contam Toxicol 51:720-729

Evans PGH (1990). European cetaceans and seabirds in an oceanographic context. Lutra 33 (2):95–12

Hammond PS, Berggren P, Benke H, Borchers DL, Collet A, Heide-Jorgensen MP, Heimlich S, Hiby AR, Leopold MF, Oien N (2002). Abundance of harbour porpoises and other cetaceans in the North Sea and adjacent waters. J Appl Ecol 39:361-376

Hiby AR (1999). The objective identification of duplicate sightings in aerial survey for porpoises. In: Garner GW, Amstrup JL, Laake JL, Manly BFJ, McDonald LL, Robertson DG (Eds.). Marine Mammal Survey and Assessment Methods. Balkema, Rotterdam, p 179-189

Hiby AR & Hammond PS (1989). Survey techniques for estimating the abundance of cetaceans. Reports of the International Whaling Commission. Special Issue 11:47-80

Hiby AR & Lovell P (1996). 1995 Baltic/North Sea Aerial surveys - Final report 11 pages and Appendix (unpubl.)

Hiby AR & Lovell P (1998.) Using airplane in tandem formation to estimate abundance of harbour porpoises. Biometrics 54:1280-1289

Huggenberger S, Benke H, Kinze CC (2002). Geographical variation in harbour porpoise (*Phocoena phocoena*) skulls: support for a separate non-migratory population in the Baltic Proper. Ophelia 56:1-12

Hutchinson J, Simmonds M, Moscrop A (1995). The harbour porpoise in the North Atlantic: a case for conservation. Conservation Research Group, University of Greenwich. London, January 1995. Report to Stitching Greenpeace Council: 90 pp.

Jepson, PD, Baker, JR, Allchin, CR, Law, RJ, Kuiken, T, Baker, JR, Rogan, E, Kirkwood, JK (1999). Investigating potential associations between chronic exposure to polychlorinated biphenyls and infectious disease mortality in harbour porpoises from England and Wales. Sci. Total Environ. 243/244, 339-348

Kaschner K (2001). Harbour porpoises in the North Sea and Baltic - bycatch and current status. Report for the Umweltstiftung WWF - Deutschland: 82 pp.

Kinze CC (1995). Exploitation of harbour porpoises (*Phocoena phocoena*) in Danish waters: a historical review. In: Bjorge A, Donovan GP (Eds.). Biology of the Phocoenids. Report of the International Whaling Commission, Special Issue 16. International Whaling Commission, Cambridge, pp. 141-153

Kiszka JJ, Haelters J, Jauniaux T (2004). The harbour porpoise (*Phocoena phocoena*) in the southern North Sea: a come-back in northern French and Belgian waters? Report for ASCOBANS. AC11/Doc. 24:1-5

Kock K-H & Benke H (1996). On the by-catch of harbour porpoise (*Phocoena phocoena)* in German fisheries in the Baltic and the North Sea. Arch Fish Mar Res 44:95-114

Koschinski S (2002). Current knowledge on harbour porpoises (*Phocoena phocoena*) in the Baltic Sea. Ophelia 55:167-197

Laake JL, Calambokidis J, Osmek SD, Rugh DJ (1997). Probability of detecting harbour porpoises from aerial surveys: estimating g(0). J Wildl Manage 61:63-75

Lockyer C & Kinze CC (2003). Status and life history of harbour porpoise, *Phocoena phocoena*, in Danish waters. NAMMCO Sci. Publ. 5:143-176

Marsh H & Sinclair DF (1989). Correcting for visibility bias in strip transect surveys of aquatic fauna. J Wildl Manage 53:1017-1024

Reijnders PJH (1992). Harbour porpoises *Phocoena phocoena* in the North Sea: Numerical responses to changes in enviromental conditions. Neth J Aquat Ecol 26:75-85

Scheidat M & Siebert U (2003). Aktueller Wissensstand zur Bewertung von anthropogenen Einflüssen auf Schweinswale in der deutschen Nordsee. Seevögel 24:50-60

Scheidat M, Kock K-H, Siebert U (2004). Summer distribution of harbour porpoise (*Phocoena phocoena*) in the German North Sea and Baltic Sea. J Cetacean Res Manage 6:251-257

Siebert U, Joiris C, Holsbeek L, Benke H, Failing K, Frese K, Petzinger E (1999). Potential relation between mercury concentrations and necropsy findings in cetaceans from German waters of the North and Baltic Seas. Mar Pollut Bull 38:285-295

Siebert U, Wünschmann A, Weiss R, Frank H, Benke H, Frese K (2001). Post-mortem findings in harbour porpoises (*Phocoena phocoena)* from the German North and Baltic Seas. J Comp Path 124:102-114

Siebert U, Gilles A, Lucke K, Ludwig M, Benke H, Kock K-H, Scheidat M (2006). A decade of harbour porpoise occurrence in German waters - Analyses of aerial surveys, incidental sightings and strandings. J Sea Res 56:65-80

Skóra KE, Pawliczka I, Klinowska M (1988). Observations of the harbour porpoise *Phocoena phocoena* on the Polish Baltic coast. Aquat Mamm 14(3):113-119

Skóra KE (1991). Notes on cetacea observed in the Polish Baltic Sea: 1979-1990. Aquat Mamm 17:67-70

Teilmann J & Lowry N (1996) Status of the harbour porpoise (*Phocoena phocoena*) in Danish waters. Rep Int Whal Commn 46: 619-625

Tiedemann R, Harder J, Gmeiner C, Haase E (1996). Mitochondrial DNA patterns of harbour porpoises (*Phocoena phocoena*) from the North and the Baltic Sea. Z Säugetierkd 61:104-111

Tiedemann R, Wiemann A, Moll K, Manteufel K (2006). Analyse der Populationsstruktur. Untersuchungen an Schweinswalen in der Ostsee als Grundlage für die Implementierung des Bestandserholungsplanes für die Schweinswale der Ostsee (Jastarnia-Plan). 2. Zwischenbericht für das Bundesamt für Naturschutz. F+E Vorhaben FKZ: 804 86 011-K1 (UFO Plan 2004): 26-39

Verfuß UK, Honnef CG, Meding A, Dähne M, Mundry R, Benke H (2007). Geographical and seasonal variation of harbour porpoise (*Phocoena phocoena*) presence in the German Baltic Sea revealed by passive acoustic monitoring. J Mar Biol Ass U.K. 87:165–176

Vinther M (1999). Bycatches of harbour porpoises (*Phocoena phocoena* L.) in Danish set-net fisheries. J Cetacean Res Manage 1:123-135

Vinther M & Larsen F (2004). Updated estimates of harbour porpoise (*Phocoena phocoena*) bycatch in the Danish North Sea bottom-set gillnet fishery. J Cetacean Res Manage 6:19-24

Wünschmann A, Siebert U, Frese K, Weiss R, Lockyer C, Heide-Jorgensen MP, Müller G, Baumgärtner W (2001). Evidence of infectious diseases in harbour porpoises (*Phocoena phocoena*) hunted in the waters of Greenland and by-caught in the German North Sea and Baltic Sea. Vet Rec 148:715-720

Excursus 2: Correlation between aerial surveys and acoustic monitoring

Ursula Siebert, Jacob Rye

Results of visual surveys and acoustic records were compared. Estimated densities (with associated confidence limits), based on aerial survey data from the Baltic Sea, were plotted for each stratum and labelled according to the month and year in which the survey had taken place (Fig. 1).

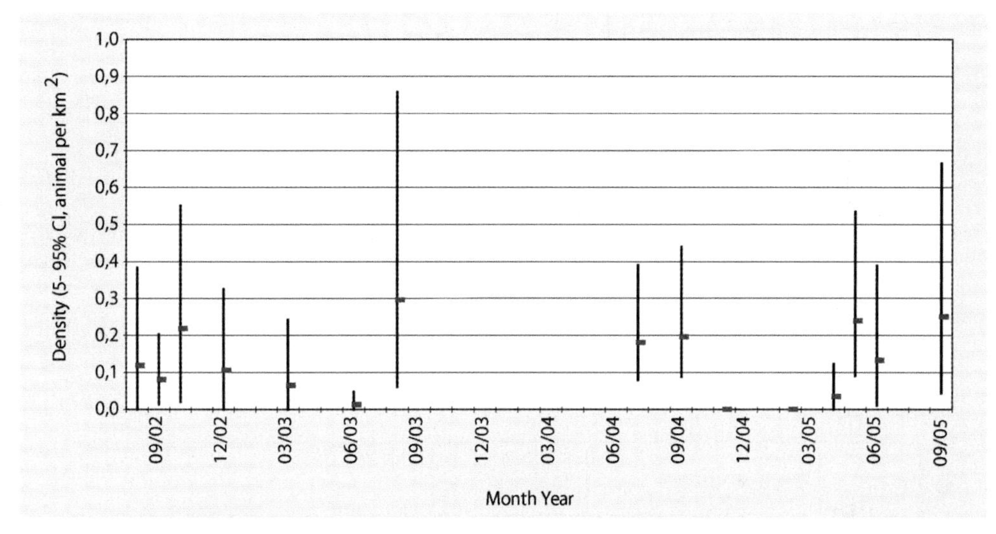

Figure 1: Estimated densities from aerial surveys in stratum F (western Baltic Sea). Associated confidence limits are indicated. In November 2004 and February 2005, surveys were conducted but no porpoises were sighted.

Chapter 2 provides details on the analysis of the aerial survey data. Similarly, the data from the T-PODs were extracted for each sub-area and month on a daily scale. The average porpoise-positive days (PPD) per month (%), the number of T-PODs in the water, the number of days with T-PODs in the water and the calculated confidence limits around the PPDs is registered. An example is shown in Figure 2.

Figure 3 depicts data from the months with both visual surveys and acoustic records in the same sub-areas.

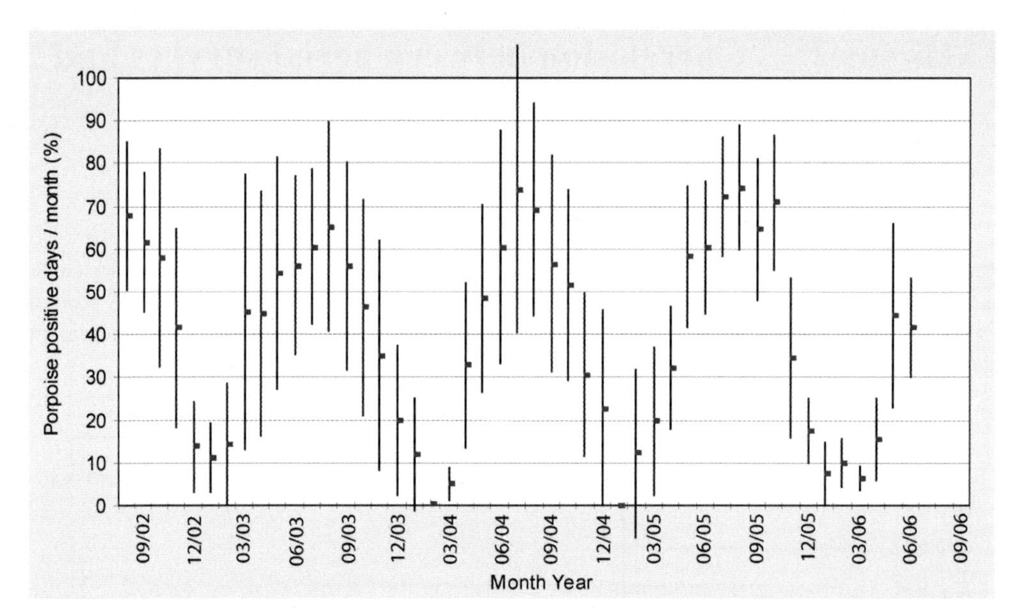

Figure 2: Percentage of porpoise-positive days per month (or %PPD/month) measured on T-POD stations in western Baltic Sea. Horizontal lines indicate the average PPD and vertical lines the associated confidence limit. E.g., the monitoring yielded an average activity of 74% in July 2004, however the confidence that the value is between 41 and 100 is only 90%. A seasonal pattern is evident, with higher values in the summer than in the winter, as explained by chapter 3.

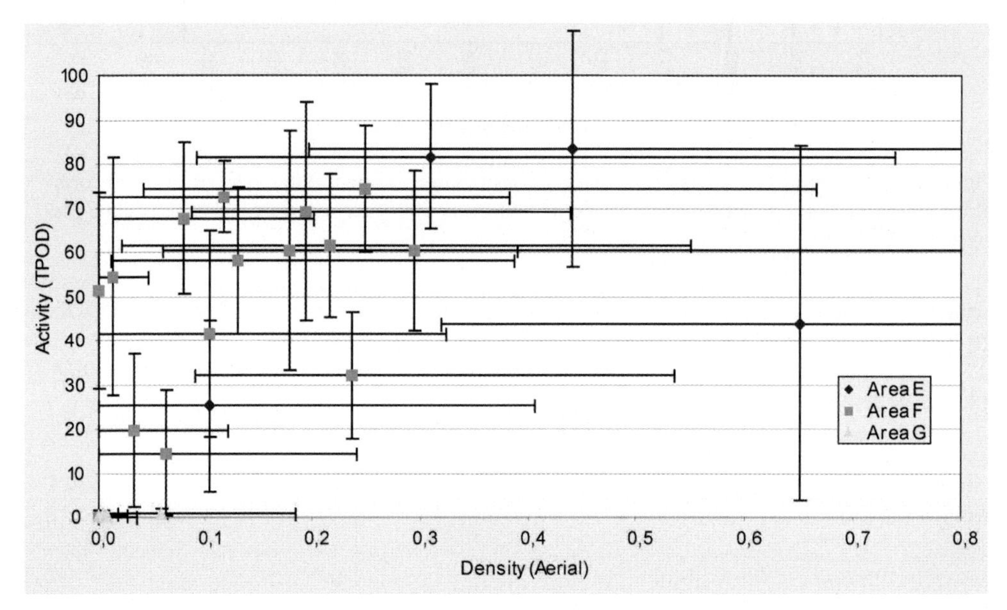

Figure 3: Comparison of visual and acoustic data in three areas of the Baltic Sea (areas defined in chapter 2). Each point represents a time period where aerial and acoustic data were available. The horizontal and vertical lines indicate the associated confidence limits for both parameters.

A correlation curve is created using a GAM analysis (Generalised Additive Model), which incorporates confidence limits (of estimated densities based on aerial survey data), number of T-PODs deployed, number of days recorded on the T-POD and area (Fig. 4).

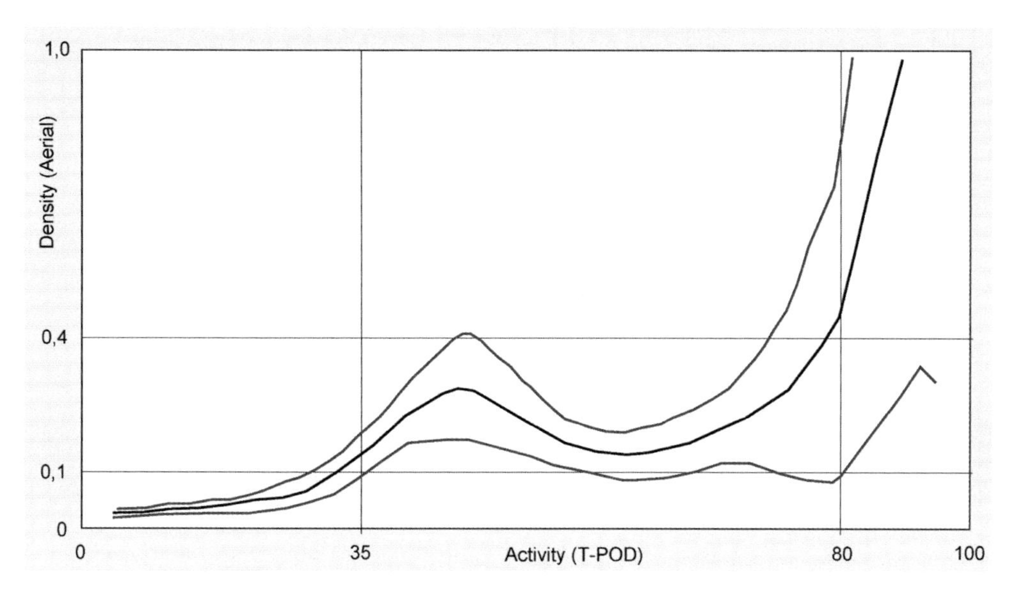

Figure 4: Correlation Curve. Points are identical to the points in Figure 2-3, axes were exchanged. The black (curved line) is the best fit, and the red lines are the 5% and 95% confidence curves.

A simplified model with a 73% fit may be created to state the following:

- If the T-PODs show detection-positive days (= porpoise-positive days) less than 35% on a monthly basis, the area has a density of less than 0.1 porpoises per km^2.
- If the detection-positive days are between 35% and 80%, there is a density of 0.1 to 0.4 porpoises per km^2.
- With detection-positive days above 80%, density is higher than 0.4 porpoises per km^2.

The grid lines in Figure 4 illustrate these densities. It should be noted that a goodness of fit of 73% is remarkable for a model that is based on so few data points (n=16). Fits as low as 30% would have been considered acceptable.

Although it may be intuitive to say that more porpoises in an area means more acoustic registrations from porpoises, this is the first time that it can be statistically confirmed, and it is one of the major findings of the intercalibration sub-project of MINOS+. The implications of this finding are also of great importance. Future work on abundance will likely still be based on visual surveys since they are specifically designed for this purpose and well tested. Further double-method studies with simultaneous aerial surveys and T-POD deployments are still needed, both to improve correlation and to obtain data in areas with harbour porpoise densities of 0.4 or higher. The current correlation is only valid for the Baltic Sea (and in the western part possibly only during some periods of the year). In areas with higher densities, such as most of the German North Sea, the time scale on the T-POD data should be shortened, e.g. to detection positive hours. Some data from the North Sea are available through MINOS and MINOS+ but they are insufficient for statistical analysis or sensible estimations.

3 The history of the German Baltic Sea harbour porpoise acoustic monitoring at the German Oceanographic Museum

Ursula K. Verfuß, Christopher G. Honnef, Anja Meding, Michael Dähne, Sven Adler, Annette Kilian, Harald Benke

Zusammenfassung

Schweinswale besitzen einen hoch entwickelten Echoortungssinn. Das so genannte Biosonar ist der Hauptsinn dieser kleinen Zahnwale. Sie produzieren eine Abfolge von sehr kurzen hochfrequenten Klicklauten und nutzen das zurückkommende Echo zur Orientierung, Navigation und Nahrungssuche. Der beinahe ununterbrochene Einsatz ihres Biosonars und ihre Mobilität macht ein passiv akustisches Monitoring dieser Walart sehr effizient.

Klickdetektoren (T-POD) erkennen die Schweinswal-Klicklaute und speichern den genauen Zeitpunkt jeder Schweinswalregistrierung. Das Deutsche Meeresmuseum (DMM) untersuchte die Einsatzmöglichkeit dieses Datenloggers für ein passiv akustisches Monitoring von Schweinswalen in der deutschen Ostsee. Weiterhin entwickelte das DMM eine Kalibriermethode, um die akustischen Eigenschaften dieser Geräte zu bestimmen. Mit einem Messnetz von bis zu 42 T-POD-Stationen entlang der deutschen Ostseeküste sowie in ausgewählten Gebieten der Ausschließlichen Wirtschaftszone (AWZ) untersuchte das DMM die deutsche Schweinswalpopulation von 2002 bis 2007 über fast fünf Jahre.

Innerhalb weniger Untersuchungsmonate zeigte sich ein geographischer Unterschied im Schweinswalvorkommen zwischen der westlichen und östlichen deutschen Ostsee. Während an den westlichen Positionen sehr viele Schweinswale registriert wurden, konnten im östlichen Teil nur selten Tiere aufgezeichnet werden. Auch saisonale Änderungen wurden während des ersten Untersuchungsjahres deutlich. Der prozentuale Anteil an Tagen mit Schweinswalregistrierungen nahm vom Frühjahr zum Sommer hin zu und wurde vom Herbst zum Winter wieder weniger. Beide Phänomene erwiesen sich über den gesamten Untersuchungszeitraum als statistisch signifikante Ergebnisse. Vergleiche mit visuellen Bestandserfassungen verdeutlichten, dass die Ergebnisse des akustischen Monitoring die relative Schweinswaldichte widerspiegeln.

Die aktuelle Studie zeigt, dass die deutsche Ostsee permanent durch Schweinswale genutzt wird, mit geographischen Unterschieden in der Dichte und einem saisonalen Wanderungsverhalten. Die erhöhte Registrierungsrate von Schweinswalen von Frühjahr bis Herbst im Vergleich zum Winter lassen die Schlussfolgerung zu, dass die deutsche Ostsee ein wichtiges Gebiet für die Vermehrung und die Aufzucht dieser Kleinwale ist.

Durch das passiv akustische Monitoring werden zeitliche und geographische Veränderungen in der Schweinswaldichte innerhalb kurzer Zeit sichtbar. Sie ist eine effektive Methode, das Vorkommen und Wanderungsverhalten von Schweinswalen zu untersuchen.

Abstract

Harbour porpoises possess an impressive and highly sophisticated echolocation sense. This so called biosonar is the major sense of these small toothed whales. They produce a series of very short high frequency click impulses and use the returning echoes for orientation, navigation and food acquisition. The nearly constant use of their biosonar and their high mobility make a static acoustic monitoring of this cetacean species very efficient.

Porpoise detectors (T-POD) recognise porpoise click trains and record the exact time of each porpoise registration. The German Oceanographic Museum (GOM) tested the benefits of these loggers for static acoustic monitoring of harbour porpoises in the German Baltic Sea. The GOM also developed a calibration procedure to define the acoustic properties of these devices. With a network of up to 42 measuring positions stationed along the coast of the German Baltic Sea as well as in selected areas of the German Exclusive Economic Zone, the GOM monitored the German harbour porpoise population for almost five years, from 2002 to 2007.

Within the first few months of monitoring, a geographical difference in porpoise presence between the western and the eastern part of the German Baltic Sea became apparent. While quite a few porpoises were registered at western positions, porpoise encounters were rare in the eastern part. Seasonal changes became obvious during the course of the first year. The percentage of days with porpoise registrations on the monitoring days increased from spring to summer, and decreased again from autumn to winter. Both phenomena recurred during the entire monitoring period and proved to be significant. Comparison with visual abundance estimates revealed that the results of the acoustic monitoring mirrors changes in harbour porpoise density.

The present study confirms year-round, regular use of the German Baltic Sea by harbour porpoises along with geographical differences in density and a seasonal migration pattern. The increased harbour porpoise detection from spring to autumn, compared to winter, leads to the conclusion that the German Baltic Sea is an important breeding and mating area for these small cetaceans.

Through static acoustic monitoring, changes in harbour porpoise density became apparent within a comparably short monitoring period on a temporal and geographical scale. This methodology proves to be a valuable and effective tool for investigating the presence of harbour porpoises and their migratory movements.

Background

Harbour porpoises possess an impressive and highly sophisticated sense, the echolocation, also called biosonar. This active sensory system developed independently twice during the evolution of mammals: in bats and in toothed whales. The species of both mammalian branches actively produce pulsed high-frequency sounds and process the returning echoes arising from sound reflections by items in the surrounding environment. Bats produce the echolocation pulses with their larynx and emit them through their mouth or nose, just like mammals produce sound. In contrast, odontocetes have developed a secondary sound source located within their nasal passages, the phonic lips (Cranford & Amundin 2004), and emit the echolocation sounds through a beam forming fat body on their forehead, the 'melon' (Goodson et al. 2004). For receiving the echoes, they developed a secondary 'outer ear' - fat bodies within and around their lower jaw conducting the sound to the middle ear (Ketten 2000). Bats, in contrast, capture sound like other mammals – with their outer ear shell.

Echolocation is the major sense of harbour porpoises. While light is not very efficient in water – it gets absorbed very quickly – sound travels a much longer distance in this dense medium. This physical fact limits the perceptual range for vision to very short distances while the range of echolocation and hearing is much longer. The acoustic senses are therefore a lot more effective under water than vision.

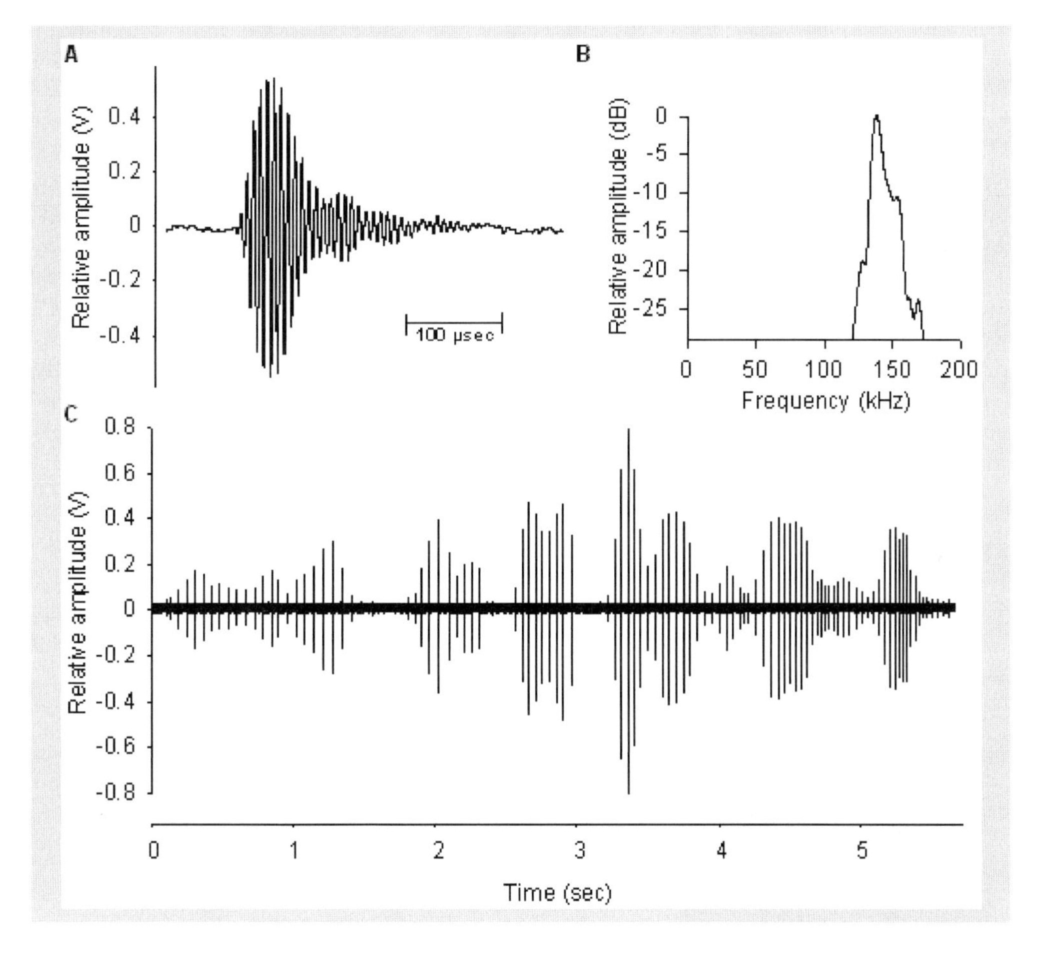

Figure 1: Amplitude-time signal (A) and power spectrum (B) of a harbour porpoise echolocation click. The amplitude-time signal graphed (C) shows an echolocation sequence built out of a series of high frequency short duration clicks as presented in (A, B).

The sound produced by harbour porpoises for echolocation is a series of very short high frequency click impulses (Fig. 1). The frequency content of such clicks lies in a small band around 130-140kHz, around seven times higher than the upper hearing threshold of humans (which is 20kHz). Those clicks are emitted in trains, with more or less regular intervals in between successive clicks. In our recordings of harbour porpoise click trains, we found click intervals from as short as 1.5ms up to 400ms and longer.

Harbour porpoises use their biosonar predominantly for navigation. When they navigate, they often focus on landmarks, prominent objects in the environment that might be used for orientation. Then they adapt their click interval to the distance to these objects (Verfuß et al. 2005). In addition to navigation, they use their active acoustic sense to find and catch fish. The echolocation click train emitted during approach and the final catch of the prey item is very distinct and typical for echolocating mammals (Verfuß & Schnitzler 2002, Verfuß et al. in prep.). While the click interval is rather long compared to how fast the echoes from the prey returns to the predator during the initial approach, the interval shortens to minimum values during the terminal phase of the catch. This sound pattern is surprisingly similar in bats and porpoises as well as other cetacean species (Schnitzler & Kalko 1998, Kalko & Schnitzler 1998, Miller et al. 1995, Madsen et al. 2005). Acoustic tags attached to the back of porpoises with suction cups showed that they nearly constantly echolocate (Akamatsu et al. 2007). When they did not echolocate or only used low amplitude clicks for echolocation (which was not distinguishable with the tag), they mostly did so over a swimming distance of less than ten meter. This makes acoustic monitoring of harbour porpoises very efficient. Additionally, porpoises are very mobile animals, swimming up to 50km per day on average (Read & Westgate 1997, Read & Gaskin 1985, Westgate et al. 1995), which is beneficial for stationary passive acoustic monitoring as the chance that a porpoise sojourning in a monitored area will eventually pass an acoustic device is high. The very high frequency and unique spectrum of the porpoises' clicks is the third reason why these animals are easy to monitor and distinguishable from dolphins whose clicks are quite different in frequency content from porpoise clicks.

Methods

Tregenza (1998) introduced a harbour porpoise detector (Fig. 2) that he and his colleagues developed for monitoring the echolocation behaviour of harbour porpoises around gill nets, as the incidental by-catch of these animals in fisher nets is one of the major threats to them.

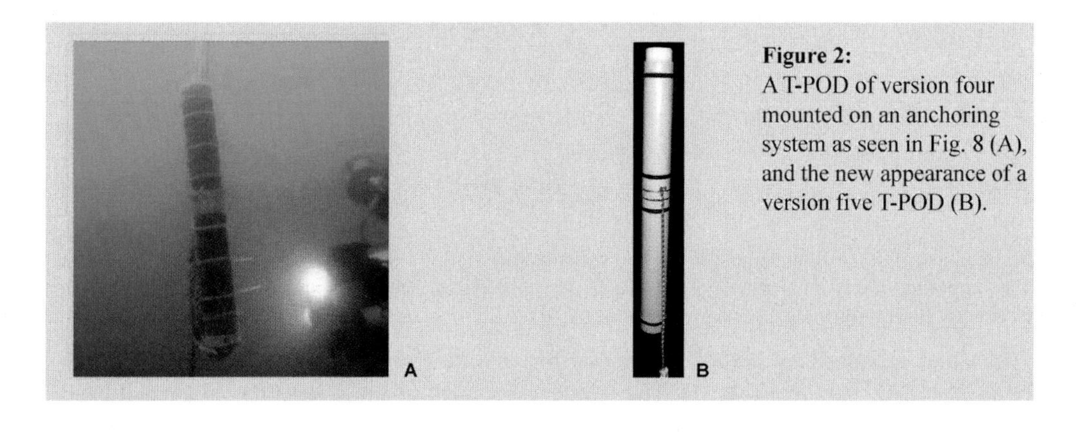

Figure 2:
A T-POD of version four mounted on an anchoring system as seen in Fig. 8 (A), and the new appearance of a version five T-POD (B).

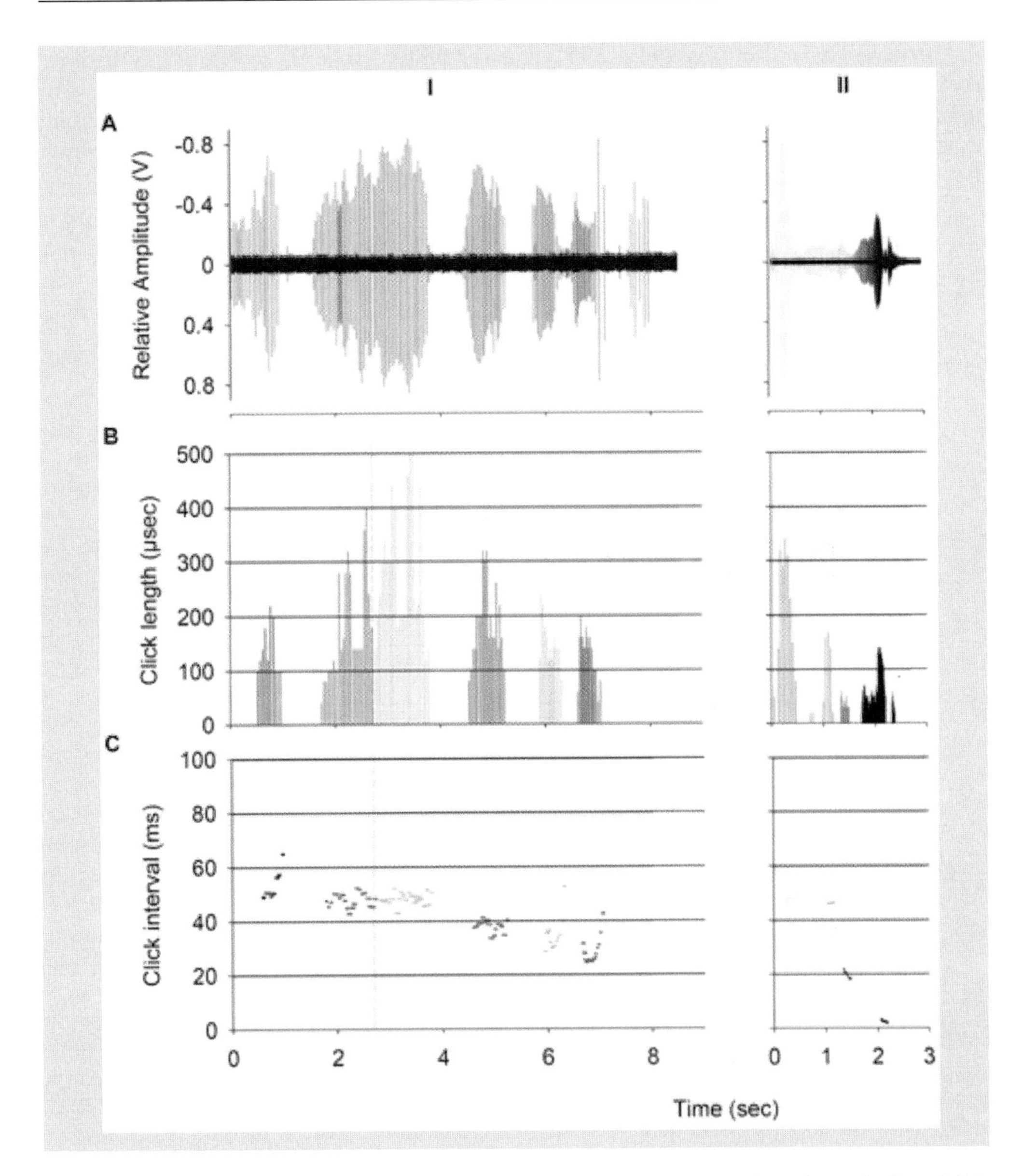

Figure 3: Amplitude-time signal (A) of a click train recorded during an orientation task (I) and fish catch (II) of a harbour porpoise in the semi-natural outdoor enclosure of the Fjord & Belt Centre, Kerteminde, Denmark (see text for details). The same echolocation sequences as given in (A) were registered simultaneously by a T-POD and classified as trains. The T-POD programme display of the length of the clicks (B) and especially the interval in between each successive click (C) can be used for interpreting the porpoise's behaviour.

In 2001, the German Oceanographic Museum (GOM) started a project, financed by the German Federal Ministry for the Environment, Nature Conservation and Nuclear Safety (BMU), to test these loggers for possible use in stationary passive acoustic monitoring of harbour porpoises in the German and Polish Baltic Sea (Verfuß et al. 2004). This project started with testing the first version of the T-POD, the **T**iming **PO**rpoise **D**etector which registers the exact time of occurrence and length of specific click events. These events need to feature a frequency spectrum matching specific criteria of harbour porpoise echolocation clicks.

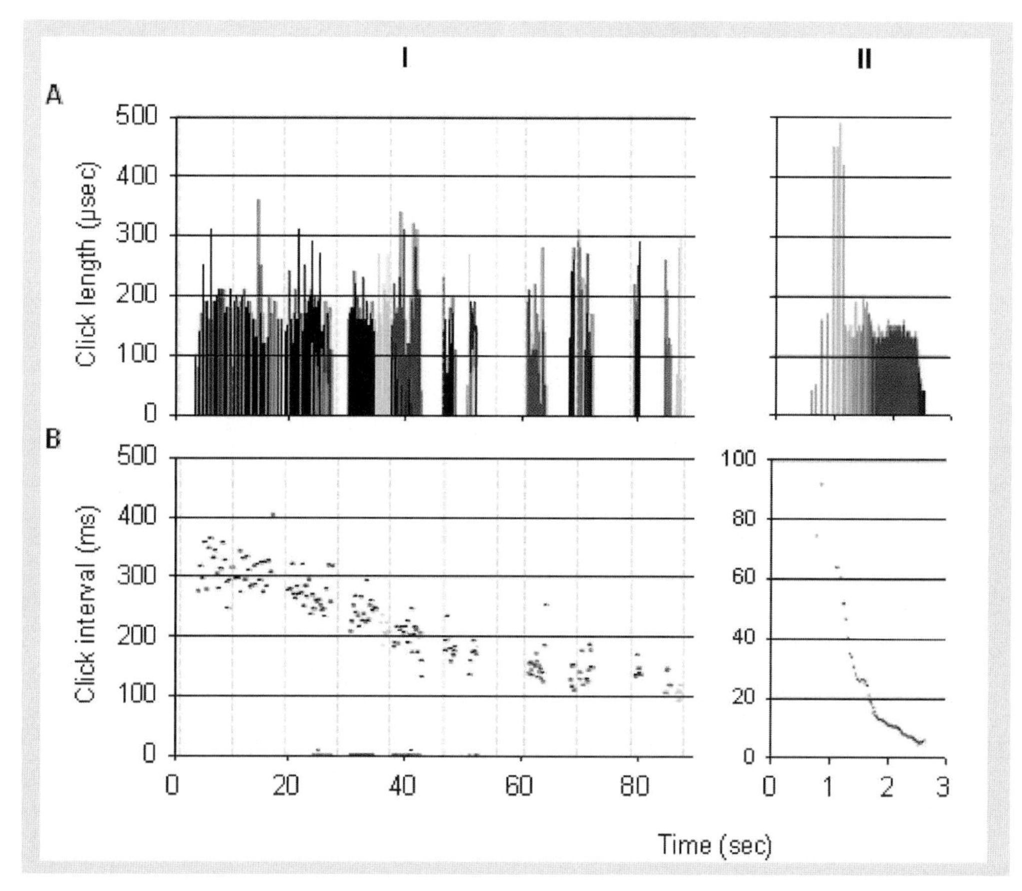

Figure 4: Two click trains of harbour porpoises as registered and classified by a T-POD in the German Baltic Sea. The graphs show the click length (A) and click interval (B) as displayed by the T-POD programme. The click pattern of train (I) and (II) are comparable to those of Figure 3 (I) and (II), respectively, and allow for an interpretation of the porpoise's behaviour.

Aside from harbour porpoises, there are other sources in the oceanic environment producing high frequency sounds which match the T-POD's criteria and therefore are registered by the logger. Those include boat sonars using frequencies above 120kHz, air bubbles bursting when high waves collapse, moving sand bottoms at stormy weather or cavitations from rotating propellers. All of these sources may cause registration on a T-POD but they are unlikely to produce click train patterns similar to those used by porpoises for echolocation. The software (T-POD.exe) for viewing and processing T-POD-data comprises a pattern recognition algorithm, searching in all registered events for sequential clicks that form a train, and classifies them, separating porpoise echolocation sequences from those of other sources.

The initial project was planned to put the T-PODs through rigorous testing (Verfuß et al. 2004). The goal was to verify whether classified click trains - said to be from porpoises - are actually of true porpoise origin. To determine whether different porpoise behaviours may be distinguished with the help of T-POD data, experiments were conducted with two captive harbour porpoises in a semi-natural outdoor pool of the Fjord & Bælt Centre in Kerteminde, Denmark. The trained porpoises were told to catch dead fish, or to swim from one side of the pool to the other, similar to the experiments conducted by Verfuß & Schnitzler 2002, Verfuß et al. 2005, and Verfuß et al. (in prep.). At the catch side or destination, respectively, T-PODs and a hydrophone were deployed to obtain simultaneous broadband sound recordings. The T-PODs recognised the porpoise click trains, the algorithm classified them as being of porpoise origin, and the registration of each single click enabled the evaluation of the porpoises' behaviour based on the click pattern (Fig. 3, Fig. 4).

The detection range of the T-PODs was investigated in the same project with simultaneous T-POD monitoring and visual observation of the monitoring area at Fyns Hoved, a nature reserve in Denmark with high cliffs and view of a porpoise-rich area. Time and position of porpoise sightings were correlated with registrations on the T-PODs and obtained a 250m minimum radius detection range for version one T-PODs, while version two T-PODs detected porpoises up to a distance of 470m (Verfuß et al. 2004).

Before beginning acoustic monitoring, it was necessary to determine whether datasets obtained by different T-PODs are comparable. Listening devices like the T-PODs have specific detection thresholds for sound, i.e. sound needs to have a certain amplitude to be detected. This implies that T-PODs with higher thresholds are less sensitive. They may register fewer clicks because they do not register sound below certain amplitudes that more sensitive T-PODs would register. To manage this problem, a calibration procedure was developed. The detection threshold of each porpoise detector was determined by sending a series of harbour porpoise clicks with decreasing amplitude at the detector in a water filled calibration pool (Fig. 5) and calculating the absolute sound pressure level of the minimum amplitude registered at the detector. This threshold value was determined for eight positions in 45 degree steps in the horizontal plane of the T-POD's receiver (Fig. 6). While the first versions of the T-PODs showed considerable individual differences in their sensitivity, the detectors became more and more similar in their acoustic properties with increasing version number. Furthermore, it became possible to raise or lower sensitivity, thus adjusting T-PODs to one standard sensitivity value (Fig. 6).

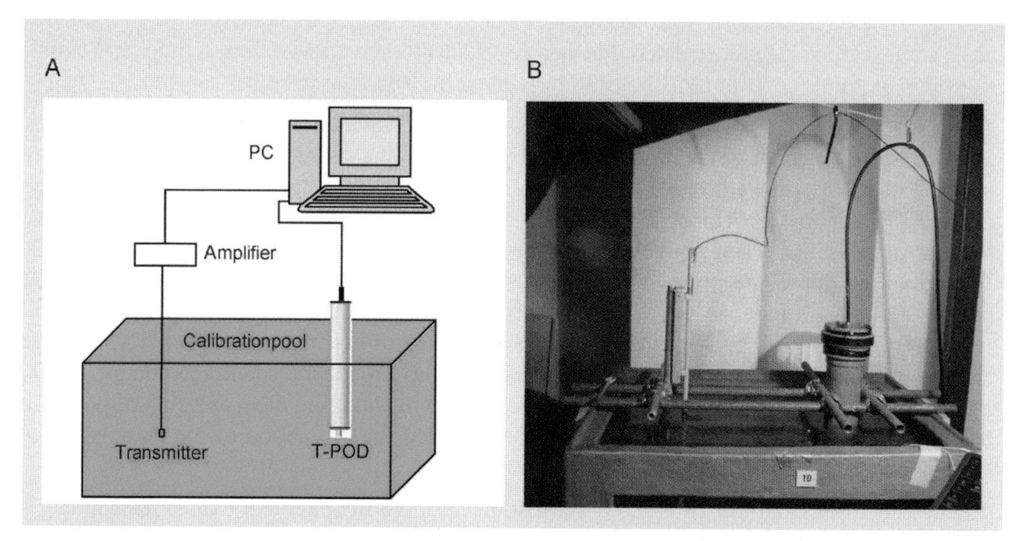

Figure 5: Schematic drawing (A) and picture (B) of the current calibration set-up. A train of harbour porpoise clicks with decreasing amplitude is transmitted into the calibration pool with the help of a PC connected to a power amplifier and transmitter. Those clicks can be received by a T-POD (seen in (A)) or a control hydrophone (seen in (B)) installed 50cm apart. The hydrophone is connected via an amplifier to the PC to record the click train. Those recordings are used to calculate the detection threshold, and therefore the sensitivity of the T-PODs.

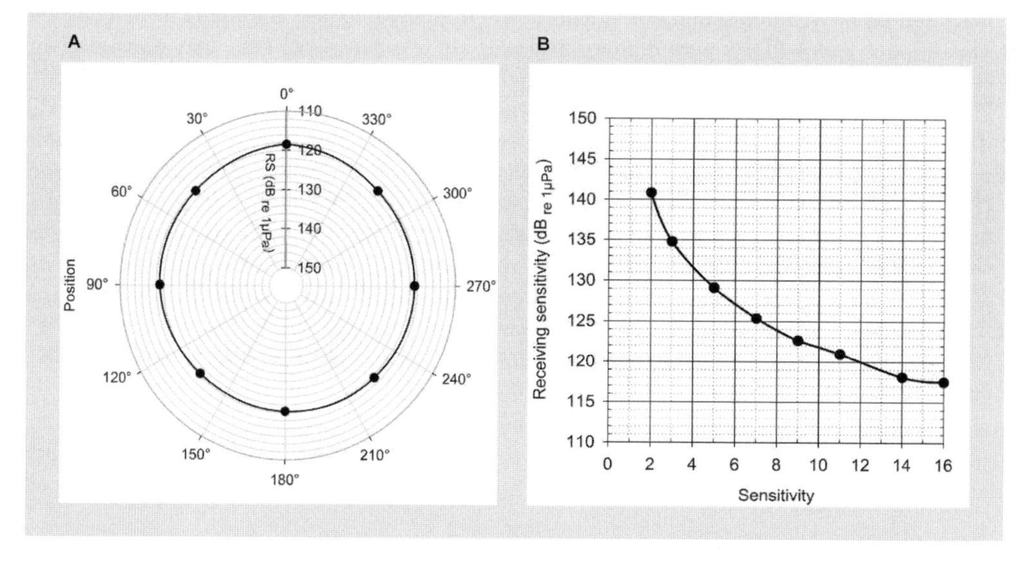

Figure 6: The graphs show the results of a calibration session: a T-POD's receiving beam pattern (A) and its receiving sensitivity (RS) (B). Both graphs give the minimum sound pressure level that porpoise clicks need for being registered by a T-POD, depending on the T-POD-hydrophone's position facing the transmitter in the horizontal plane (see Fig. 5) (A) and on the T-POD setting 'sensitivity' (B).

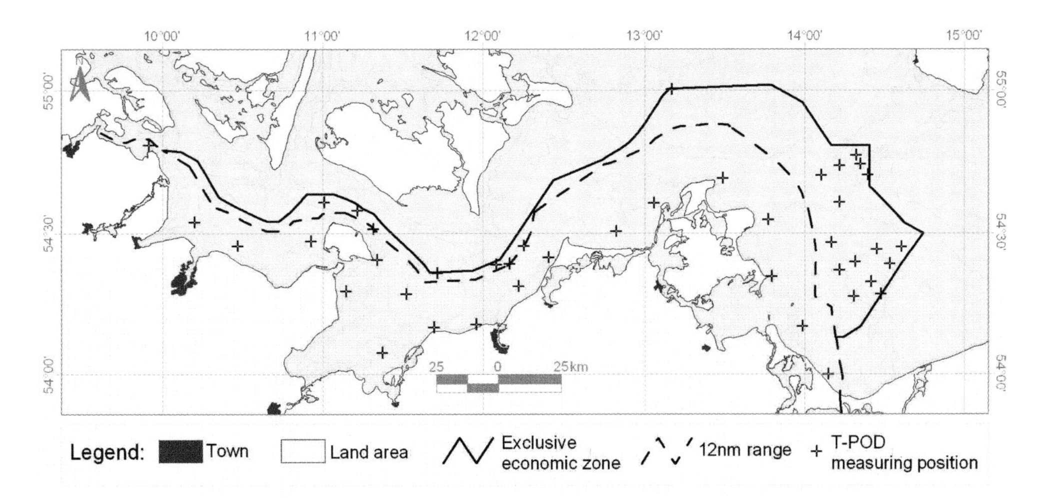

Figure 7: T-POD measuring positions (crosses) throughout the German Baltic Sea.

After it had been confirmed that the T-PODs were able to detect the presence of harbour porpoises in a specific area and that any differences in sensitivity could be managed statistically, planning began for a net of measuring positions stationed along the coast of the German Baltic Sea as well as in selected areas of the German Exclusive Economic Zone, for instance the Fehmarnbelt, the Kadet Trench, Kriegersflak and the Pomeranian Bay (Fig. 7).

This measuring net was operated by cooperating GOM projects financed by the Federal Ministry of the Environment and the Federal Agency for Nature Conservation. It was extended into the Kiel and Mecklenburg Bight after MINOS with the start of MINOS+ and another cooperating project that had a special interest in the Pomeranian Bay harbour porpoise stock. It was possible to monitor the German Baltic Sea for almost five years from 2002 to 2007 with up to 42 measuring positions. T-PODs were moored on lightweight anchoring systems (Fig. 8) maintainable from small Zodiac boats. Maintenance services were scheduled every eight to twelve weeks to check the anchoring system, download the data, and change the batteries.

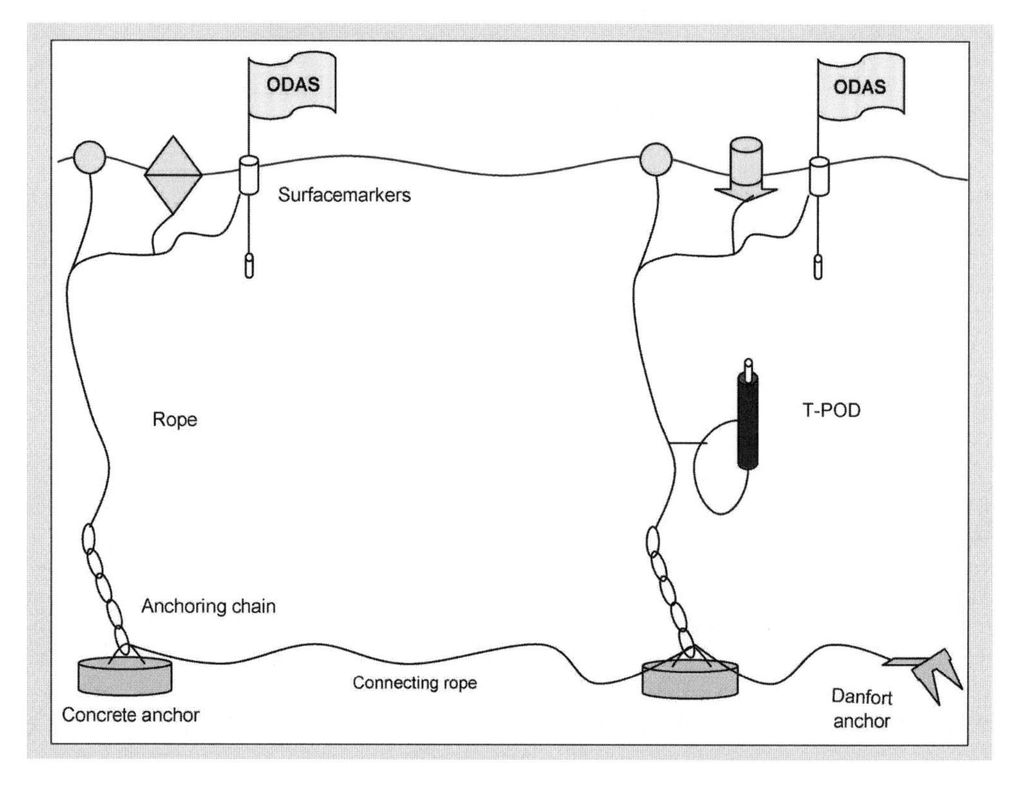

Figure 8: Schematic drawing of the T-POD anchoring system as used by the German Oceanographic Museum for the measuring positions shown in Fig. 7

Results

Within the first few months of the static acoustic monitoring, a geographical difference in porpoise presence between the western and the eastern part of the German Baltic Sea became apparent. While quite a few porpoises were registered at western positions, porpoise encounters were rare in the eastern part. The low registration rate of porpoises in the eastern German Baltic Sea made it necessary to not only rely on the classified porpoise trains but also to visually inspect all other click trains that were classified to be from sources other than porpoises, although they may have originated from these animals. Most of these classifications were from other sources and unlikely from porpoises. However, some porpoise click trains were found in these classes and as every porpoise registration was important in this low porpoise density area, visual inspection was applied to all retrieved data.

In addition to the geographical difference in porpoise presence, a seasonal change became apparent during the course of the first year. The percentage of days with porpoise registrations on the monitoring days - % porpoise positive day (%PPD) - increased from spring to summer, and decreased again from autumn to winter. Both phenomena recurred during the whole monitoring period and proved to be significant (Fig. 9) (Verfuß et al. 2007). Comparison of %PPD with harbour porpoise density obtained for the same area and time (chapter 2, this volume) shows that changes in the %PPD mirrors changes in the harbour porpoise density and that they positively correlate (see Fig. 3 in Excursus 2). The seasonal changes of %PPD speaks for a migration of harbour porpoises, most likely from the Danish Belt Sea, into the western German Baltic Sea in spring, a spreading eastward during summer, and a westerly migration during autumn and out of the German Baltic Sea in the winter months (Fig. 9). It should be noted that porpoises are present year round in these German waters, though with different densities and density distributions.

Until the mid-20th century it was believed that harbour porpoises migrated into and out of the Baltic Sea (reviewed in Koschinski 2002), although it has never been scientifically proven. It was assumed that porpoises follow the movements of herring in spring through Danish waters into the Baltic Sea. In late autumn and winter, when the Baltic tended to freeze over in some years, the porpoises may have migrated back out of the Baltic Sea. Nowadays, the porpoise population is very small (see chapter 2) and migration is difficult to prove. Teilmann et al. (2004) were able to prove seasonality in the use of areas in Danish waters with the help of satellite tags on porpoises. Siebert et al. (2006) also showed seasonality of incidental sightings and strandings in the German Baltic Sea with a maximum of sightings and strandings in the summer months July to September, based on data from more than ten years. The authors discuss the possibility that the data from incidental sightings may be biased by a lower effort in winter (e.g. fewer sailing boats), whereas the strandings data, obtained by a year-round observer scheme and standard procedure ensuring stable monitoring efforts, may be biased by a longer submersion time of carcasses when water temperature is low (Moreno et al. 1993) and the unknown drift route that the dead animals may take.

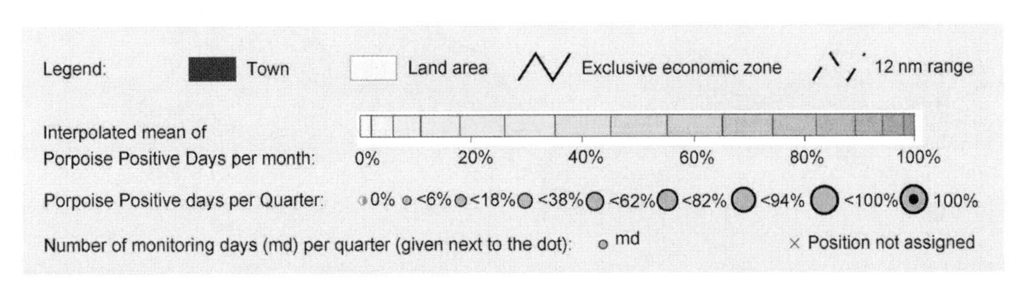

Figure 9: (Legend on this page, graphs on the following pages) Results of the static acoustic monitoring of the years 2005 (A) and 2006 (B). Shown are the percentages of porpoise positive days for the quarters of the year (I: Jan – Mar, II: Apr – Jun, III: Jul – Sep, IV: Oct – Dec). The size of the grey circles represents the percentage of porpoise positive days per quarter as obtained by the corresponding measuring stations. The number of monitoring days is given next to the circles. Modelling of the data with the help of a GAM-statistics visualise nicely the yearly repeating seasonal and geographical differences in harbour porpoise density by colour coding the interpolated mean percentage of porpoise positive days per month.

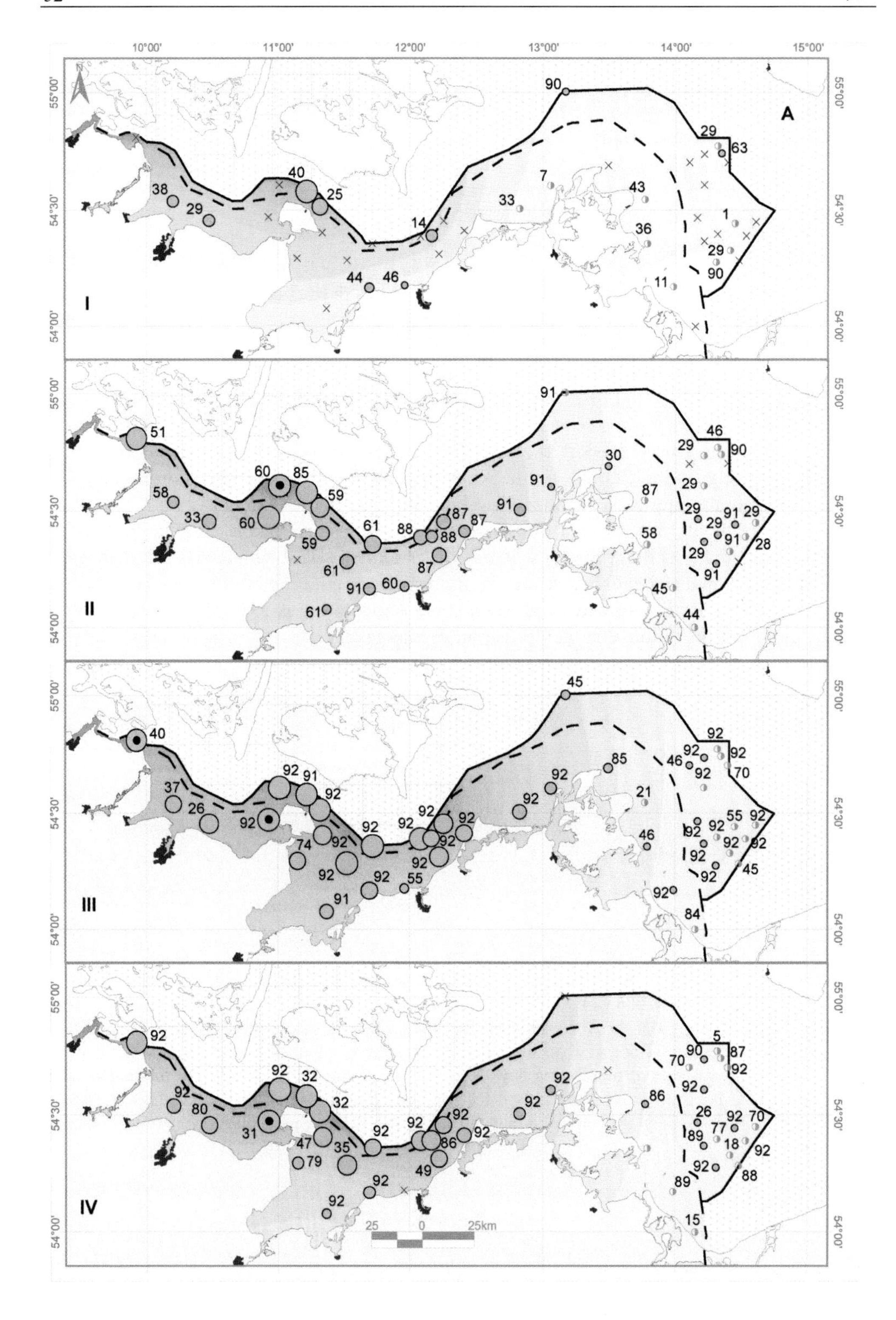
10°00'
11°00'
12°00'
13°00'
14°00'
15°00'
55°00'
54°30'
54°00'
A
I
II
III
IV
25
0
25km

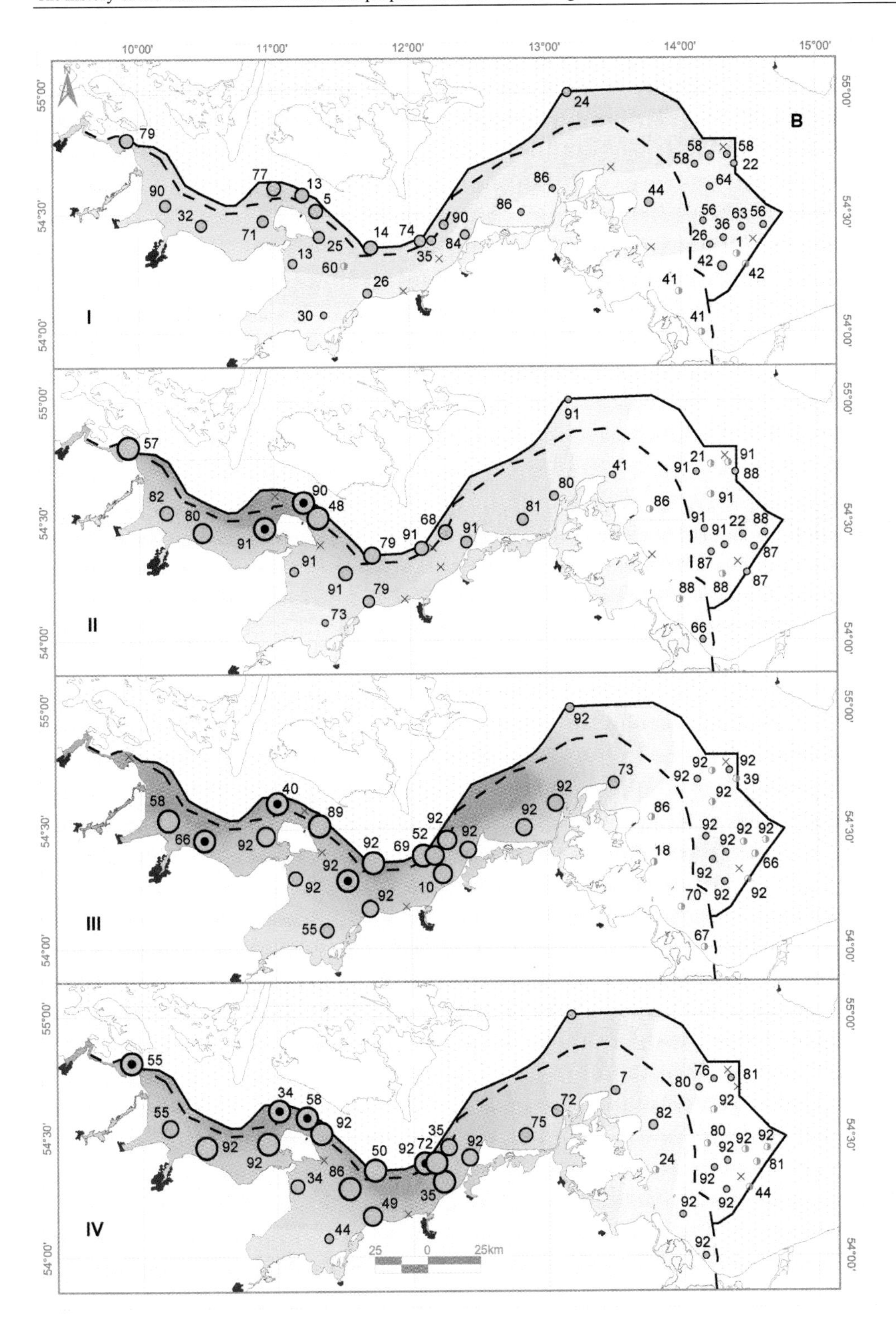
10°00'
11°00'
12°00'
13°00'
14°00'
15°00'
55°00'
54°30'
54°00'
N
B
I
II
III
IV
25
0
25km

Through static acoustic monitoring, changes in harbour porpoise density became apparent on a temporal and geographical scale within a comparably short monitoring period. This methodology proves to be a valuable and effective tool for investigating the presence of harbour porpoises and their migratory movements. The present study confirms regular use of the German Baltic Sea by harbour porpoises, along with geographical differences in density and a seasonal migration pattern. The increased harbour porpoise detection from spring to autumn, compared to winter, leads to the conclusion that the German Baltic Sea is an important breeding and mating area for harbour porpoises.

Acknowlegdements

The research was funded by the German Federal Ministry for the Environment, Nature Conservation and Nuclear Safety (BMU), including the Investment-in-future program (ZIP) as part of the projects MINOS and MINOS+, and by the Federal Agency for Nature Conservation (BfN). These data could not have been obtained without the administrative help of the Federal Maritime and Hydrographic Agency, the Coastguard Service, the Federal Border Guard, the Water and Shipping Authorities Stralsund and Lübeck. We are grateful for the cooperation, assistance and discussions given by Dr. Nick Tregenza, and the support of the Fjord & Bælt Centre staff. We would like to thank the students, scientific helpers and volunteers. The following individuals contributed significantly to data acquisition and analysis: Ines Baresel, Anja Brandecker, Heiko Charwat, Jakob Rye Hansen, Martin Jabbusch, Marcel Klehm, Susanne Kotzian, Kathrin Krügel, Annika Krutwa, Peter Leopold, Roger Mundry as well as the crews of the Seefuchs and Palaemon.

References

Akamatsu T, Teilmann J, Miller LA, Tougaard J, Dietz R, Wang D, Wang K, Siebert U, Naito Y (2007). Comparison of echolocation behaviour between coastal and riverine porpoises. Deep-Sea Research Part II, 54:290-297.

Cranford TW & Amundin M (2004). Biosonar pulse production in odontocetes: The state of our knowledge. In: Echolocation in Bats and Dolphins. Thomas JA, Moss CF, Vater M (Eds.), pp. 27-35. Chicago: The University of Chicago Press.

Goodson AD, Flint JA, Cranford TW (2004). The harbour porpoise (*Phocoena phocoena*): Modelling the sonar transmission mechanism. In: Echolocation in Bats and Dolphins. Thomas JA, Moss CF, Vater M (Eds.), pp. 64-85. Chicago: The University of Chicago Press.

Kalko EKV & Schnitzler H-U (1998). How echolocating bats approach and acquire food. In: Bat Biology and Conservation. Kunz TH & Racey PA (Eds.), pp. 197-204. Washington: Smithsonian Institution Press.

Ketten DR (2000). Cetacean ears. In: Hearing by whales and dolphins. Au WWL, Popper AN, Fay RR (Eds.), pp. 43-108. New York: Springer.

Koschinski S (2002). Current knowledge on harbour porpoises (*Phocoena phocoena*) in the Baltic Sea. Ophelia, 55(3):167-197.

Madsen PT, Johnson M, Aguilar de Soto N, Zimmer WMX, Tyack PL (2005). Biosonar performance of foraging beaked whales (*Mesoplodon densirostris*). The Journal of Experimental Biology, 208:181-194.

Miller LA, Pristed J, Møhl B, Surlykke A (1995). The click-sounds of narwhals (*Monodon monoceros*) in Inglefield Bay, northwest Greenland. Marine Mammal Science, 11(4):491-502.

Moreno P, Benke H, Lutter S (1993). Behaviour of harbour porpoise (*Phocoena phocoena*) carcasses in the German Bight: surfacing rate, decomposition and drift routes. Untersuchungen über Bestand, Gesundheitszustand und Wanderung der Kleinwalpopulationen (Cetacea) in deutschen Gewässern. Institut für Haustierkunde, Universität Kiel, Germany, FKZ 108 050 17/11, BMU-final report.

Read AJ & Westgate AJ (1997). Monitoring the movements of harbour porpoises (*Phocoena phocoena*) with satellite telemetry. Mar.Biol., 130:315-322.

Read AJ & Gaskin DE (1985). Radio tracking the movements and activities of harbour porpoises, *Phocoena phocoena* (L.), in the Bay of Fundy, Canada. Fishery Bulletin, 83:543-552.

Schnitzler H-U & Kalko EKV (1998). How echolocating bats search and find food. In: Bat Biology and Conservation, pp. 183-196. Washington: Smithsonian Institution Press.

Siebert U, Gilles A, Lucke K, Ludwig M, Benke H, Kock K-H, Scheidat M (2006). A decade of harbour porpoise occurrence in German waters - Analyses of aerial surveys, incidental sightings and strandings. Journal of Sea Research, 56:65-80.

Teilmann J, Dietz R, Larsen F, Desportes G, Geertsen BM, Andersen LW, Aastrup P, Hansen JR, Buholzer L (2004). Satellitsporing af marsvin i danske og tilstødendefarvande. Danmarks Miljøundersøgelser, Electronic Version: http://www2.dmu.dk/1_viden/2_Publikationer/3_fagraporter/abstrakter/abs_484_DK.asp, DMU 484, 86 pp.

Tregenza NJ (1998). Site acoustic monitoring for cetaceans - a self-contained sonar click detector. In: Proceedings of the Seismic and Marine mammals Workshop. Tasker ML & Weir C (Eds.), pp. 1-5. London: Sea Mammal Research Unit.

Verfuß UK & Schnitzler H-U (2002). F+E Vorhaben: Untersuchungen zum Echoortungsverhalten der Schweinswale (*Phocoena phocoena*) als Grundlage für Schutzmaßnahmen. Eberhard Karls Universität Tübingen, Lehrstuhl Tierphysiologie, Tübingen, FKZ 898 86 021, final report.

Verfuß UK, Honnef CG, Benke H (2004). Untersuchungen zur Nutzung ausgewählter Gebiete der Deutschen und Polnischen Ostsee durch Schweinswale mit Hilfe akustischer Methoden. FKZ 901 86 020, final report.

Verfuß UK, Miller LA, Schnitzler HU (2005). Spatial orientation in echolocating harbour porpoises (*Phocoena phocoena*). The Journal of Experimental Biology, 208:3385-3394.

Verfuß UK, Honnef CG., Meding A, Dähne M, Mundry R, Benke H (2007). Geographical and seasonal variation of harbour porpoise (*Phocoena phocoena*) presence in the German Baltic Sea revealed by passive acoustic monitoring. Journal of the Marine Biological Association of the United Kingdom, 87:165-176.

Verfuß UK, Miller LA, Pilz PK, Schnitzler H-U (in prep). Echolocation in foraging harbour porpoises (*Phocoena phocoena*).

Westgate AJ, Read AJ, Berggren P, Koopmann HN, Gaskin DE (1995). Diving behaviour of harbour porpoises, *Phocoena phocoena*. Canadian Journal of Fisheries and Aquatic Sciences, 52(5):1064-1073.

Excursus 3: Harbour porpoises in the North Sea - Tidal dependency

Ursula Siebert, Jacob Rye

One of the first results from the operation of T-PODs in tidal areas was the discovery of differences in acoustic registrations recorded by T-PODs as a function of tidal rhythm at a station in Meldorf Bight, as shown in an example in Figure 1: The tidal state is indicated by the small lines at the top of the picture, showing the average angle of the T-POD's suspension during each 30 min period. Vertical lines indicate high or low tides, with low water current.

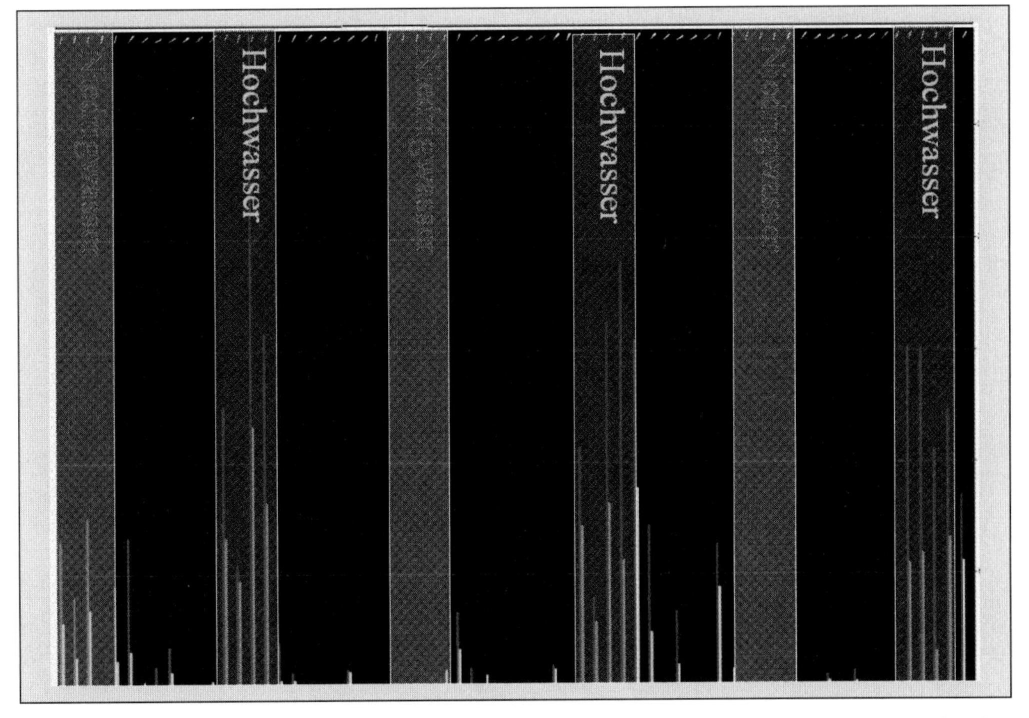

Figure 1: Periodic rhythm in porpoise registrations. Screenshot from the T-POD programme. The size of the bar indicates the number of clicks assigned to either high probability (red) or low probability (yellow) porpoise echolocation click trains per 30 minutes.

From the time stamp on the file we can assign the three aggregations of porpoise registrations to periods with high tide. The data files are much longer than the 33 hours shown above, and statistical analysis with 2-tailed T-tests showed significantly more porpoise-positive minutes (%) per tidal state (high, low, running tide 1 or 2) at high tide (HT) compared with low tide (LT) and running tide (FT) after a high tide (Fig. 2 A).

Tidal currents in the bay, at times exceeding 5 knots, seemed to be responsible for the observed phenomenon. In other areas, such as around the Horns Reef wind farm in the Danish North Sea and in an area in the Irish Sea, harbour porpoise registrations were shown to be higher with stronger currents (Frank Thomsen and Henrik Skov, Pers. Com.) and at flood tides. Better food availability at increased water speeds was believed to have caused a higher porpoise density because porpoises can simply hold their position and let food pass by them. The absence (or reduced registration) of porpoises at the same tidal periods in Meldorf Bight may be explained with the very high current speed which prevents porpoises from holding their position in the water for extended periods. Instead, they would have to wait for slower moving water and then come in and feed on the prey brought in by the water flowing into the bay.

Further investigation however showed that this conclusion is not sustainable. Figure 2 B was created with data from the same station 2 months later in the year. The same statistics were applied. Now, the only statistical difference is between low tide and the two running tides, with more registrations at low tide. The biological mechanism for this change is unknown. It should be noted that the overall porpoise registration is lower in the summer period, compared to the spring data in the previous figures.

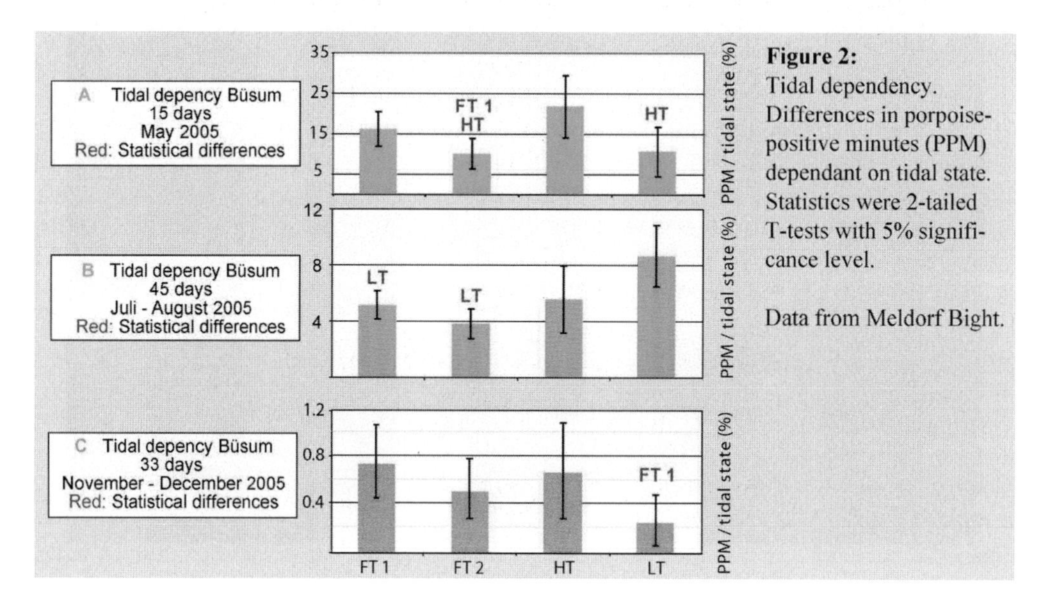

Figure 2: Tidal dependency. Differences in porpoise-positive minutes (PPM) dependant on tidal state. Statistics were 2-tailed T-tests with 5% significance level.

Data from Meldorf Bight.

Data from the winter (Fig. 2 C) showed even lower overall registrations, but here the low tide has the lower value in the number of registrations compared to the running tide following the low tide.

In conclusion, it appears that, while it is possible to find a correlation between tidal state and harbour porpoise registrations, it is not the tidal state itself but something else that controls the presence of harbour porpoises in this area. Further investigations are needed to understand this phenomenon.

4 How tolerant are harbour porpoises to underwater sound?

Klaus Lucke, Paul A. Lepper, Marie-Anne Blanchet, Ursula Siebert

Zusammenfassung

Die geplante Errichtung von Offshore-Windenergieanlagen (WEAs) im deutschen Bereich der Nord- und Ostsee wird mit der wiederholten Erzeugung intensiver Schallemissionen einhergehen, wenn die Fundamente der Anlagen in den Boden gerammt werden. Die Rammstöße werden im Durchschnitt einen Schalldruckpegel von 225 dB re 1µPa in 1 m übersteigen. Schweinswale sind die einzige Walart, die in den deutschen Gewässern der Nord- und Ostsee heimisch sind. Diese Tiere sind aufgrund ihrer Lebensweise vollständig auf ein gesundes Gehör angewiesen. Bisherige Untersuchungen an Delfinen und Belugas geben Anlass zu der Vermutung, dass eine Beschallung mit derartigen Rammstößen bei Schweinswalen zumindest zu einer zeitweiligen Beeinträchtigung des Gehörs (TTS) führen kann. Um einen Beschallungsgrenzwert für diese Art ableiten zu können, wurde eine audiometrische Studie an einem Schweinswal durchgeführt.

Die Untersuchung erfolgte an einem im Fjord- & Bælt Centre in Kerteminde, Dänemark gehaltenen Schweinswal. Die Hörfähigkeit des Tieres wurde durch Messung akustisch evozierter Potenzial (AEP-Methode) ermittelt. Nachdem ein Audiogramm für das Tier erstellt worden war, wurde der Schweinswal in einem kontrollierten Beschallungsexperiment mit Schallimpulsen einer Airgun beschallt. Nach jeder Beschallung wurde bei drei Frequenzen die Hörempfindlichkeit des Tieres auf signifikante Änderungen hin überprüft. Der empfangene Schalldruckpegel der Airgun-Impulse wurde solange gesteigert, bis TTS bei mindestens einer Frequenz nachgewiesen werden konnte.

Das zunächst ermittelte Audiogramm des Schweinswals weist im Vergleich zu veröffentlichten Daten von anderen Tieren erhöhte Hörschwellenwerte auf. Die erhöhten Hörschwellenwerte können durch eine elektrophysiologische Maskierung der gemessenen AEPs, die akustische Maskierung aufgrund des hohen Hintergrundschallpegels im Untersuchungsbereich und die akustischen Charakteristika der AEP-Stimuli bedingt sein. Entsprechend müssen die gemessenen Werte als maskierte Hörschwellenwerte angesehen werden. Dies hat jedoch keine Auswirkungen auf die Untersuchungen zur Toleranz des Gehörs für intensive Schallimpulse.

Nach einer Beschallung mit einem maximalen Empfangspegel von 200 dB re 1 µPa und einer Energieflussdichte von 164 dB re 1 $\mu Pa^2/Hz$ konnte bei dem Tier bei 4 kHz erstmals TTS nachgewiesen werden. Die ermittelte TTS-Schwelle des Schweinswals liegt deutlich niedriger als vergleichbare Werte anderer Zahnwalarten. Es ist anzunehmen, dass ein größenabhängiger Zusammenhang in der akustischen Toleranz des Gehörs bei Zahnwalen besteht. Eine Modellierung der kumulativen Auswirkungen einer wiederholten Beschallung zeigt, dass der Gefährdungsbereich für Schweinswale im Umfeld einer Rammung noch größer sein wird, als auf Grund der vorliegenden Daten anzunehmen wäre.

Diese Ergebnisse sind von Bedeutung im Zusammenhang mit der Genehmigung des Baus von Offshore-WEAs in deutschen Gewässern sowie für den Einsatz vergleichbarer Schallquellen.

Abstract

The planned construction of offshore wind turbines in the North and Baltic Seas involves the emission of high numbers of intense impulsive sounds when turbine foundations are driven into the ground by pile driving. Based on knowledge about other toothed whales, it can be assumed that the source levels, which will on average exceed 225dB re 1µPa at 1m, will create a risk of at least temporary threshold shift (TTS) in the auditory system of harbour porpoises which inhabit these waters. These animals are vitally dependent on their hearing system. An auditory study was conducted to base the definition of noise exposure criteria on information on the acoustic tolerance of this species to single impulses.

The measurements were conducted on a male harbour porpoise held at the Fjord- & Bælt Centre in Kerteminde, Denmark, in a semi-natural enclosure. All hearing data were collected by measuring the evoked auditory potentials (AEP method). After achieving baseline hearing data over the animal's functional hearing range, the animal was subsequently exposed to single airgun stimuli at increasing received levels in a controlled exposure experiment. Immediately after each exposure the animal's hearing threshold was tested again for any significant changes at three selected frequencies. The received levels of the airgun impulses were increased until TTS was reached at one of the frequencies.

The animal's hearing thresholds were elevated in comparison to published data from other studies. A systematic electrophysiological masking due to the active positioning of the animal at its underwater station and an acoustic masking due to the high background noise level in the enclosure are likely reasons for these elevated hearing thresholds. The acoustic characteristics of the auditory stimuli may also account for a systematic difference in the hearing sensitivity.

The achieved harbour porpoise's hearing sensitivity does therefore not represent absolute but masked hearing threshold levels. This, however, has no implication on the tolerance of the animal's hearing system for intense impulsive sounds.

At 4kHz the TTS-criterion was exceeded when the animal was exposed to an impulse at a received peak level of 200dB re 1µPa and a sound exposure level of 164dB re 1µPa^2/Hz. The documented masked TTS level of the harbour porpoise is considerably lower than levels found in other toothed whale species tested so far, thus supporting the hypothesis of size-dependant differences in the tolerance of the auditory system in toothed whales. Modelling the impact range of multiple exposures reveals a risk for auditory effects in harbour porpoises over larger distances as compared to single exposures. The results are likely to have implications for regulatory procedures regarding the construction of offshore wind turbines in German waters as well as the use of other impulsive sound sources.

Background

The marine environment is not as quiet as one might think. Numerous sound sources exist underwater which constantly create a natural background noise of varying intensity and composition. The most prominent physical sound source are waves and rain, while snapping shrimp or singing whales are among the loudest or best known biological sound producers. A new category of sounds is introduced into the sea by the increasing use of the seas by humans. Anthropogenic sounds resulting from shipping, industrial and military activities and many other sources have led to a substantial increase in the overall background noise in the oceans over the past decades. The North and Baltic Seas are among the most intensively used and consequently also the noisiest marine areas. Currently, another large-scale anthropogenic activity is envisaged for these areas. Large numbers of wind turbines are planned to be built in offshore areas of the North and Baltic Seas as part of the efforts to increase the use of renewable energies and thus meet the goals of the Kyoto protocol and its successors. While this effort per se would be regarded as very positive, especially from an environmental point of view, its execution nevertheless raises concerns about potential adverse effects on the marine environment. One of the key issues in this respect is the noise emitted during construction of the turbines. As these structures will be built in water depths of up to 4m and will stretch to over 120m above sea level, they have to rest on very reliable structures in the ground. Current designs include foundations consisting of single, or 'Mono'-piles and those resting on three piles, the so-called Tri-Pods. Any of these piles need to be driven to a depth of 25m and more into the ground, and most often this will be achieved by using impulsive pile driving. The emissions produced during this process reach intensities with a potential of causing a variety of effects in the marine fauna, from behavioural reactions and stress to physical damage and injury. The published source levels emitted during pile driving for a wind turbine reach from 225 to 246dB re 1µPa at 1m (Ødegaard & Danneskiold-Samsøe A/S 2000; Nedwell et al. 2003, Robinson et al. 2007).

One research topic of the project MINOS+ was to elucidate the potential impact these acoustic emissions may have on marine mammals, as this group of species tend to be most sensitive to sound. Marine mammals are represented in German waters by two pinniped species, the harbour seal (*Phoca vitulina*) and the grey seal (*Halichoerus grypus*), as well as the harbour porpoise (*Phocoena phocoena*), a small toothed whale species. These three species are known or suspected to have a very acute sense of hearing underwater and, for the seals, also in air. They are also known to produce a variety of sounds. Harbour porpoises have been shown to use sounds not only for communication but also to find their prey, to orient underwater and to avoid obstacles. Especially in harbour porpoises, the acoustic sense has evolved to be their dominant sense on which they rely vitally. Any impairment or damage to their auditory system would have deleterious consequences for the affected individuals and could also lead to effects at the population level.

Problem

It is known from auditory studies on terrestrial animals that the exposure to intense impulsive sounds, comparable to the ramming impulses emitted during the construction process, could exceed the tolerance of their auditory system and lead to hearing loss. In principle, noise induced hearing loss encompasses temporary as well as permanent reduction in the auditory sensitivity of a subject. The former is defined as temporary threshold shift (TTS) if the hearing sensitivity recovers to its normal values within 30 days after the sound exposure, while all reductions in hearing sensitivity remaining after this period are defined as permanent threshold shift (PTS). TTS has been accepted as the main auditory criterion for assessing the tolerance of a hearing system to noise.

Evidence exists that the same principles apply for marine mammals with regard to the auditory effects, but when the MINOS+ studies were initiated, those data were scarce and limited to three cetacean species which are ecologically as well as from an acoustic point of view clearly different from the harbour porpoise. No information existed on the tolerance of the auditory system of harbour porpoises, and only two animals had been tested for hearing sensitivity. Immediately after the MINOS+ studies had ended, a first dose-response function and, derived from this, generalised noise exposure criteria were agreed upon based on the TTS data from the three species tested so far (Southall et al., in press). Nevertheless, a transfer of the dose-response function to the harbour porpoise and application of the noise exposure criteria to the construction of wind turbines in the North and Baltic Seas seemed questionable and would require caution.

The data obtained from bottlenose dolphins (*Tursiops truncatus*), belugas (*Delphinapterus leucas*) and a false killer whale (*Pseudorca crassidens*) indicated that a criterion combining peak pressure and energy flux density (i.e. the acoustic energy over time, SEL) of a signal should be used to predict the maximum tolerance of the auditory system of toothed whales for noise. The duration of the sound exposure is another important factor with regard to the auditory effects, as the TTS limit drops by 3dB per doubling of exposure duration. This was taken into account by differentiating between exposure to intermittent (i.e. impulsive) noise compared to continuous noise. The corresponding noise exposure criteria so far have been a peak pressure of 224dB re 1μPa and an SEL of 183dB re 1$\mu Pa^2 s$ for continuous noise, while the equivalent values for intermittent exposures are 224dB re 1μPa and a SEL of 195dB respectively (Marine Mammal Commission 2004).

Aim of research

Consequently, the aim of the acoustic study on harbour porpoises was to conduct a TTS study on a harbour porpoise and define the tolerance limit of the auditory system of the harbour porpoise to impulsive sounds. While the initial tests to establish a method that would also be applicable to wild animals – either by-caught or stranded – and the proof of concept were accomplished within the first MINOS project, the MINOS+ study concentrated on the definition of noise criteria for impulsive sound. Such data would enable regulatory agencies to define safety zones around the construction sites which would have to be cleared of harbour porpoises before construction could take place. At the same time, such data would provide reliable information to construction companies regarding the implementation of appropriate mitigation measures if their activities still exceed the noise criteria outside the safety zone.

Study design

A key element for the planned study was access to a harbour porpoise trained to participate in experiments so that the experiments could be conducted under controlled conditions and definitive information on the dose-response function gathered. Auditory tests in a controlled exposure experiment on free-ranging animals were not an option because definitive measurements of the auditory effects would not have been possible under these circumstances. The two approaches which are in principle available to conduct auditory studies on unrestrained animals are the behavioural test, as it has been widely used in psychoacoustic studies, and the electrophysiological method, which has been applied on many terrestrial animals for hearing tests. As the planned study comprised measurements of the hearing threshold of a harbour porpoise, and as these measurements should ideally be repeated on wild animals, the electrophysiological method of measuring auditory evoked potentials (AEPs) in the animals had to be chosen because this allowed in principle a rapid measurement of an audiogram, i.e. within approximately one hour.

The study was divided into two modules: The first consisted of measurements of the animal's absolute hearing thresholds over almost its entire functional frequency spectrum, thus providing a baseline for the second module, a tolerance test of the animal's hearing. This TTS test was designed to follow the same procedural structure as the experiments conducted by Finneran et al. (2002). The absolute hearing threshold was measured in half octave steps over 5.5 octaves with the lower frequency limit set by the methodological parameters of the AEP stimulation. The threshold measurements were repeated several times at three select frequencies (representing the low, mid and high frequencies of its functional hearing range) to achieve their normal variation which would subsequently allow to define a frequency-specific TTS criterion. The tolerance of the animal's auditory system would then be tested by exposing the animal to an intense sound impulse and measuring the hearing threshold again immediately afterwards. Any reduction in the animal's hearing sensitivity exceeding the preset TTS criterion would be regarded as evidence of an actual threshold shift. Subsequent measurements of the animal's hearing threshold at the affected frequency would provide information about the recovery function of the auditory system. All three frequencies would be tested separately for TTS at a given exposure level; i.e. only one hearing frequency would be tested after each exposure. As long as the hearing threshold is shown to remain within its normal variation at all three frequencies, the subsequent exposure level would be elevated and this procedure repeated until a threshold shift is detected. This careful approach was chosen to avoid any risk of permanent hearing loss.

AEP method

This AEP method is non-invasive. For this reason the technique has been widely adopted in human patients and is also used for screening of newborns.

This technique is based on the presentation of acoustic stimuli, which will generate neuronal potentials in the acoustic system upon perception of these stimuli. Two surface electrodes have to be placed on the animal's skin using suction cups – one near the blowhole and the other near the dorsal fin – to record the neural responses evoked within the auditory system. These potentials are generated within neuronal nuclei at different positions in the auditory system, thereby forming an electric field which can be detected and recorded even on the skin surface. AEPs are useful for measuring the functioning of the auditory system and examining important aspects of auditory processing. To filter out these comparatively small electric potentials from the overall neuronal, i.e. electric activity of the animal's musculature, other sensory inputs, etc., the acoustic test stimuli have to be presented at a high repetition rate. By averaging the evoked potentials (average over 500 AEPs), all non-acoustic neuronal signals can be attenuated or eliminated.

A refined methodological approach is based on the use of rhythmic sound modulations. By sinusoidally modulating the amplitude of carrier tone or sound pulse sequence, it is possible to elicit a neuronal response which includes a specific frequency component that is equivalent to the modulation frequency used. This effect occurs because the auditory system is capable of following the envelope of a sinusoidal signal and producing corresponding neuronal potentials, called Envelope-Following Response (EFR). By applying a Fast-Fourier Transformation (FFT) analysis, the modulation frequency component can be identified and quantified. The resulting amplitude of the EFR represents the energy content of the neuronal response at the given modulation frequency. Nevertheless, the strength of this EFR response can simultaneously be taken as a relative measure for the perception of the carrier frequency of the amplitude-modulated (AM-) signal. For example, no EFR response could be elicited if the signal was outside the functional and dynamic ranges of the auditory system. At each frequency, the stimuli were presented in decreasing intensity starting at a clearly audible level until a (neuronal) response was no longer detected. The resulting data were statistically tested for significance by using a F-test to identify EFR responses from arbitrarily occurring noise at the given AM frequency.

A male harbour porpoise held under human care in the Fjord- & Bælt Centre (F&B) in Kerteminde, Denmark, was chosen as subject for the studies. This animal, named *Eigil*, was estimated to be between 9 and 10 years old in 2005 when the study began. A comprehensive medical record of all treatments exists for *Eigil* for almost its entire live since he was brought to this facility at the age of 1. He was held in this facility with two female harbour porpoises at that time. The older female was pregnant twice during the study period from 2005 until 2007 and gave birth to a female calf right after the end of the studies in summer 2007. The pregnancies interfered with the design of the auditory experiments and thus are relevant for discussion of the results.

The animals are kept at the F&B in a semi-natural outdoor pool of 30m by 20m and an average depth of 4m. Their enclosure stretches along the entrance from the Baltic Sea to a small Fjord on one side of the busy fishing harbour of Kerteminde. It has a natural bottom and solid walls of concrete and steel on the two long sides. It is separated from the harbour on its narrow ends by nets, thereby providing a constant water exchange with the Baltic Sea. The enclosure is divided into two compartments ('main pool' and 'research pool'), allowing separation of the animals for experiments. A floating pen (4.5m x 4.5m x 1.5m) in the research pool was wrapped with sound absorbing foam, providing an acoustic shelter for the two females during the planned exposures to intense sound in the later stage of the study.

All experiments were conducted with *Eigil*, who was separated from the two females to avoid behavioural or acoustic interference between the animals during the research. *Eigil* was trained to accept the electrodes being attached to its head and back with suction cups and to dive on command to an underwater station in 1.5m water depth. The training method used was based on operant conditioning and positive reinforcement. No food deprivation was used during these experiments. He stationed himself actively at this station with his rostrum touching a 4 x 4cm PVC plate in front of the sound transducers (Fig. 1). He stayed there for 100s on average until he was called back to the surface by the trainer to receive a reinforcement. This experimental sequence was called a 'send'. A complete research session was comprised of 4 sends on average. The number of research sessions per day depended on weather conditions and varied between 1 and 4 during the study period with an average of 2 sessions, ideally one in the morning and one in the afternoon.

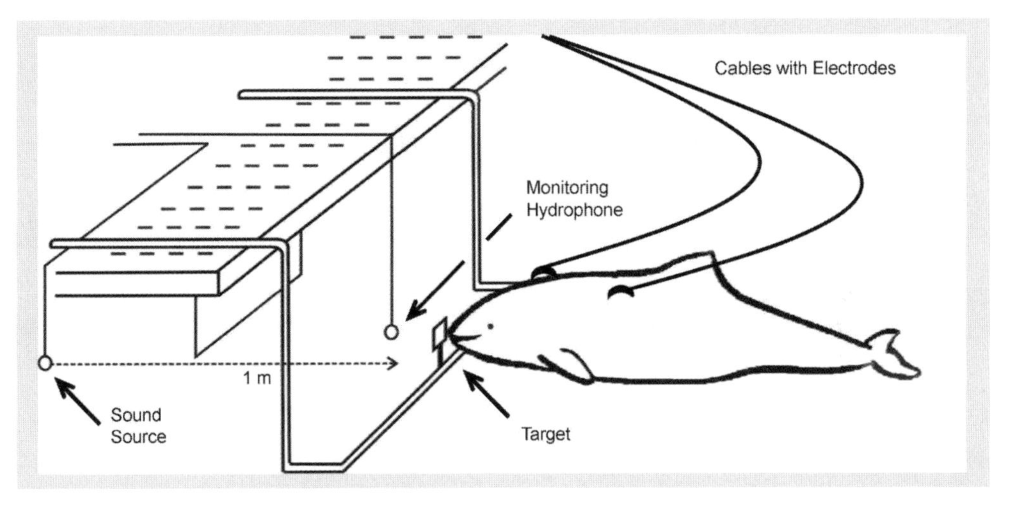

Figure 1: Schematic plot of the research setup for the AEP measurements with the animal positioning itself at 1.5m water depth in front of its underwater station and with its body in a straight line with the sound path of the incoming AEP stimuli.

The animals' hearing was tested at frequencies between 4kHz and 160kHz. A custom-made software application was used to programme all acoustic stimuli that were transmitted to elicit the AEPs during the hearing threshold tests. The signal generation system consisted of a data acquisition card (National Instruments DAQ 6062 E) and two function generators (Thurlby Thandar TG 230 and Agilent 33220A – with the first triggering the latter). The signals were amplitude-modulated sine waves of 25ms duration with a modulation depth of factor 1 and a cosine envelope. At frequencies between 4kHz and 8kHz all signals were amplified by a power amplifier PA 100E (Ling Dynamic Systems Ltd., Royston, UK) and transmitted via an underwater transducer USRD J-9.

At higher frequencies a power amplifier Brüel & Kjaer 2713 was used to amplify the signals. Due to differences in their transmit response, four different sound transducers were used to transmit the acoustic stimuli during the AEP tests: a Reson TC 4033 at 16 and 80kHz, a SRD 4" ball hydrophone at 22.4kHz, a SRD HS70 at 44.8kHz and a SRD 150 at the remaining frequencies. All transmitted and received signals were constantly observed in real time at an oscilloscope and recorded for post-hoc analysis via a monitoring hydrophone (Reson TC 4014) and a preamplifier (etec B1501) for received level, signal quality and undesired signal artefacts. The evoked potentials were fed into a custom-built input station consisting of an amplifier (20dB gain) and an optical separation unit (including 20dB gain). Additionally, the signals were bandpass filtered (highpass frequency: 300Hz, lowpass frequency: 10kHz, NF

Electronic instruments FV-665) to avoid artefacts. Each sequence of one hundred successive potentials was averaged and displayed online as well as stored for post-hoc analysis.

The background noise in Kerteminde harbour is dominated by shipping noise from a variety of boat traffic ranging from recreational boats and small fishing boats passing the enclosure to fishing boats turning into the unloading area on the opposite side of the harbour or a supply vessel for a nearby island. The background noise was thus dominated by low frequency noise at varying levels and frequencies, depending on the size, speed and activity of the respective boats.

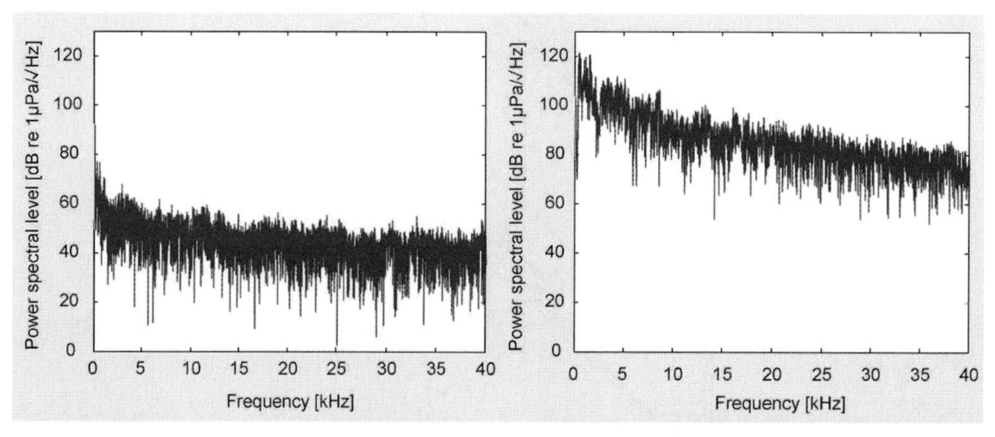

Figure 2: Background noise level (plotted as pressure spectral density) recorded in the research pool at F&B during quiet conditions.

Figure 3: Background noise level (plotted as pressure spectral density) recorded in the research pool at F&B recorded at the same position as in Figure 2 during noisy conditions (motorboat turning at 50m distance).

A small sleeve airgun (20in^3) was used as sound source to produce the fatiguing sound stimuli during the second module (Fig. 4). This device was pressurised with Nitrogen at a pressure of 137 bar (2000 PSI) and was operated at a depth of 2m (i.e. in mid-water) from an small boat (source boat) in Kerteminde harbour at varying positions between the F&B and the eastern exit of the harbour area. The exact position of the source boat was determined by GPS and this information along with time, weather conditions and other relevant information on the sound source were documented for further analysis. An intensive calibration of the airgun had been conducted prior to the study to predict the received levels of the airgun stimuli as a function of its distance to the receiving position in the main pool at the F&B. The airgun was shot at distances between 2km (in Kerteminde Bay, outside the harbour area) and 15m from the receiving position at the F&B. All animals were kept in the sound-insulated floating pen during these experiments and none of them showed any behavioural reaction to the airgun stimuli. None of the airgun signals could be detected in the monitoring recordings conducted inside the floating pen except the one with the device shot nearest to the F&B and thereby resulting in the loudest airgun signals.

Figure 4: Picture of the airgun used as a sound source during the TTS experiments. The device was suspended from a rotatable crane and lowered into the water over a winch system.

Based on these data a predictive 2-D sound propagation model (Fig. 5) was calculated, which provided sufficient accuracy for the determination of the initial distance of the source boat from the F&B.

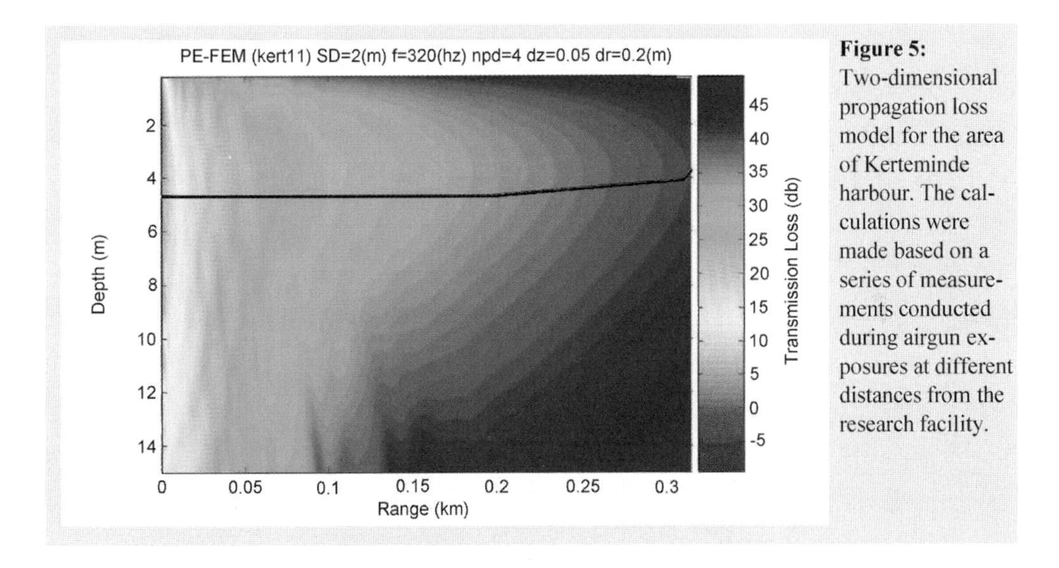

Figure 5: Two-dimensional propagation loss model for the area of Kerteminde harbour. The calculations were made based on a series of measurements conducted during airgun exposures at different distances from the research facility.

The sudden release of pressure from the airgun during a 'shot' results in an oscillating air bubble which projects a short, intense impulse (Fig. 6) into the water and across adjacent boundaries (ground wave; audibility of airgun shot in air). The main acoustic energy of this impulse is centred below 500Hz but considerable energy can also be detected up to above 20kHz, well above low levels of background noise (Fig. 7).

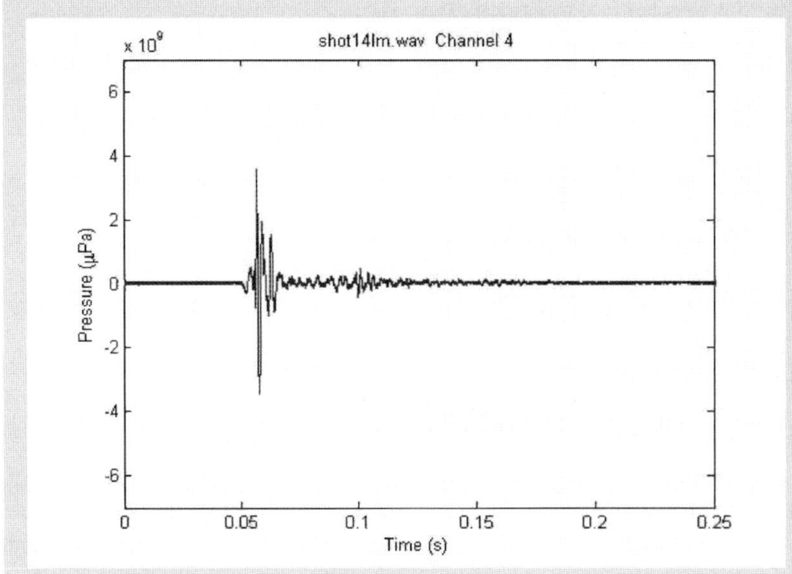

Figure 6: Time domain representation of an airgun impulse. The airgun was fired at 2m water depth in Kerteminde harbour and the impulse was recorded at a distance of 15m to the receiving hydrophone.

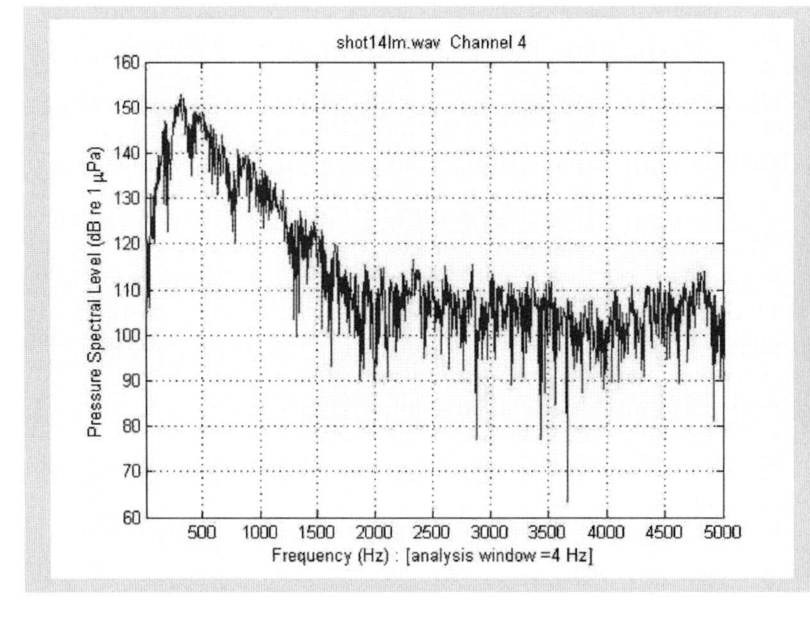

Figure 7: Frequency spectrum analysis of the recorded airgun impulse (Figure 6) showing the pressure spectral level [dB re 1μPa]. The frequency spectrum is plotted in Hz, the spectrum levels are based on a 4Hz analysis band.

Prior to every airgun shot, the two female harbour porpoises were separated into the sound-insulated floating pen. Their general behaviour and breathing rate were observed for the period of the sound exposure and compared with baseline data previously obtained under normal conditions. *Eigil* remained in the main pool. A receiving hydrophone was positioned at 1.5m water depth at a position at the narrow end of the pool facing towards the eastern exit of

Kerteminde harbour. This position had proven to receive the most intense signals during the airgun calibration. The airgun was triggered as soon as *Eigil* was within 2m of the receiving hydrophone with its body fully underwater. His behaviour was monitored and recorded for further analysis.

Immediately after each exposure to the fatiguing stimulus, *Eigil* was then led into the research pool where the AEP setup was located. Within 4min after the exposure the AEP measurements were started and lasted 12min on average. Within this period his hearing sensitivity could be determined at a single frequency.

During the second module, *Eigil*'s hearing sensitivity was tested at 4kHz, 32kHz and 100kHz, thus representing the low, mid and high frequency range of the functional hearing spectrum of the harbour porpoise.

Results

Eigil's hearing threshold was determined at frequencies between 4kHz and 140kHz. At the highest frequency tested, 160kHz, no threshold could be obtained at all. The measurements of *Eigil*'s auditory sensitivity at the remaining frequencies resulted in much higher thresholds compared to hearing data published for other harbour porpoises.

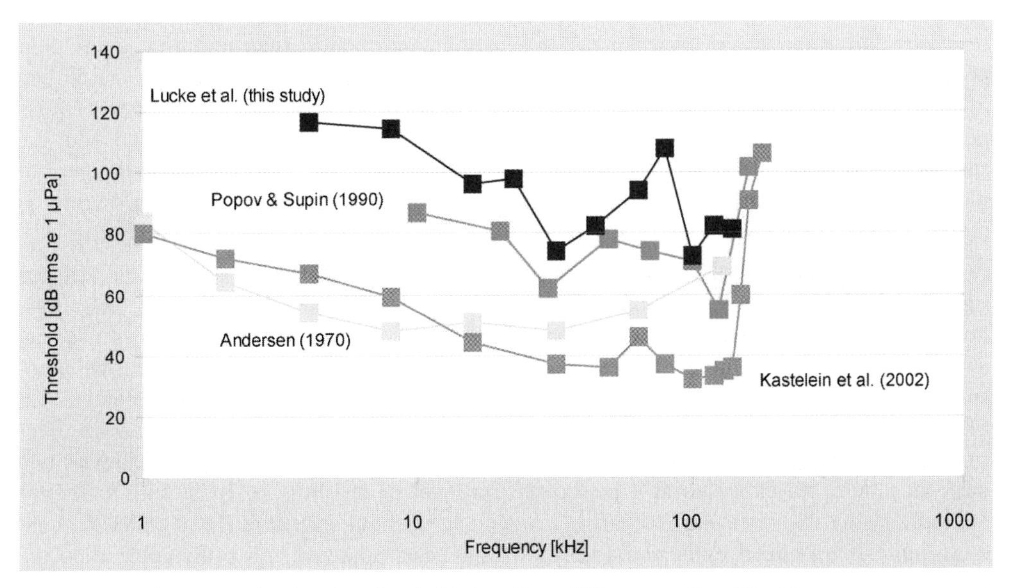

Figure 8: Harbour porpoise hearing threshold data from different studies. The blue squares represent the threshold values achieved in this study. Data from another AEP study (Popov & Supin 1990) as well as from two behavioural auditory studies (Andersen 1970, Kastelein et al. 2002) are given for comparison.

The shape of *Eigil*'s hearing curve with its two minima at the mid and high frequency range is in good accordance with the previously published data. However, a clear rise in threshold of 50dB on average was measured compared to data obtained by Kastelein et al. (2002) in a behavioural hearing study, with a maximum difference at 80kHz. At the higher frequencies, the difference is not as pronounced compared to the thresholds obtained by Andersen (1970) in the first behavioural hearing study conducted on harbour porpoises. Compared to the results from the AEP study by Popov & Supin (1990), *Eigil*'s thresholds are elevated by roughly 10dB.

Over a period of 4.5 month *Eigil* was exposed to a total of 24 airgun impulses. The received peak pressure (peak to peak) of the pulses ranged from 161.2dB re 1µPa to 202.2dB re 1µPa, with an acoustic energy (SEL) ranging from 140.5dB re 1µPa^2s to 167.2dB re 1µPa^2s. Those levels were reached at distances between 150m and 14m from the animal's position during the exposure. *Eigil* nevertheless showed no behavioural reactions during the first exposures when he was exposed to a received peak pressure level of 174dB re 1 µPa or an SEL of 145dB re 1µPa^2s. At received levels above, *Eigil* showed a typical aversive reaction at the time of the sound exposure and behavioural avoidance towards the location of the shot reactions and subsequently avoided approaching the exposure station. After a temporary threshold shift had been documented and confirmed, the received levels were not raised any higher and no further trials were conducted.

Because one of the female harbour porpoises was pregnant during the exposure period, special measures were taken to protect her and the other animal from unnecessary sound exposures. Both females were kept in a sound-insulated pool and their behaviour was continuously monitored during the sound exposures. None of them showed any obvious behavioural reactions during the airgun experiments. The attenuation of the airgun impulses inside their pool was at the order of 30 to 40dB. Moreover, this floating pen was installed at the far end of the research pool, hence more than 25m away from the monitoring hydrophone at *Eigil*'s position during the exposures. The sound attenuation due to the foam, combined with the spreading loss over the increased distance, led to a reduced received level inside the pool of more than 40dB. Accordingly, the two females were never exposed to peak pressure levels of more than 160 B re 1µPa.

A temporary threshold shift was first measured after *Eigil* had been exposed to an airgun impulse at a peak pressure of 200.2dB re 1µPa and a SEL of 164.5dB re 1µPa^2s. The threshold shift was measured when *Eigil*'s hearing was tested after the exposure for its sensitivity at 4kHz. Since this threshold shift was only 1.8dB above the preset TTS criterion, the exposure was repeated a few days later with a received peak pressure level of 202.1dB and a SEL of 165.5dB re 1µPa^2s. The resulting threshold shift at 4kHz was 9.1dB above the TTS criterion and hence a clear proof of TTS. Another verification of this effect was achieved two days later, after an exposure at a peak pressure level of 201.9dB re 1µPa with a SEL of 165.8dB re 1µPa^2s, when *Eigil*'s hearing revealed a threshold shift at 4kHz of 15dB. A slight elevation of *Eigil*'s hearing threshold at 32kHz had been measured at a comparable exposure level before, and no change in hearing sensitivity was measured in *Eigil* after an exposure to those levels at 100kHz.

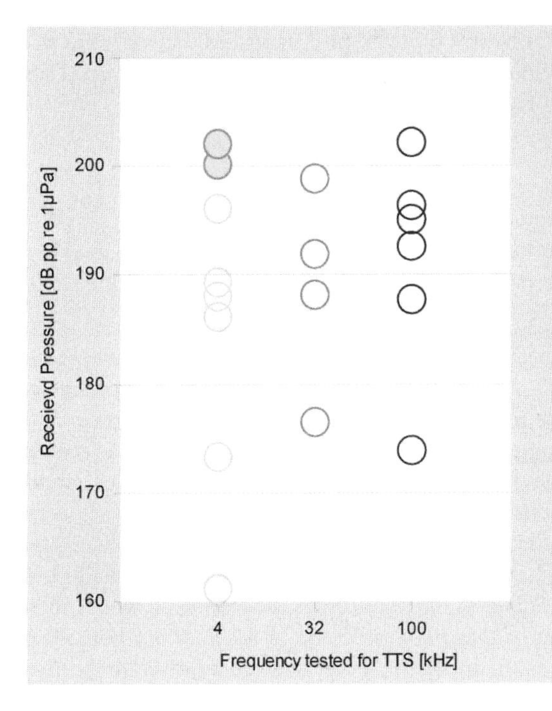

Figure 9: Effect of exposure to different levels of fatiguing sound stimuli on Eigil's hearing thresholds tested at three different frequencies. Open circles represent post-exposure hearing thresholds within the normal variation of threshold values, filled circles show post-exposure thresholds above the normal variation, thus representing a threshold shift.

Recovery

An important factor for the assessment of this noise-induced effect is the recovery of *Eigil*'s auditory system. After the first clear threshold shift had been measured, a series of AEP measurements was conducted over the next days to follow the further development of *Eigil*'s hearing sensitivity at the affected frequency. 178 minutes after the initial exposure *Eigil*'s hearing had recovered only partially from its threshold shift, as it was reduced by 2.9dB but still being elevated above the TTS criterion.

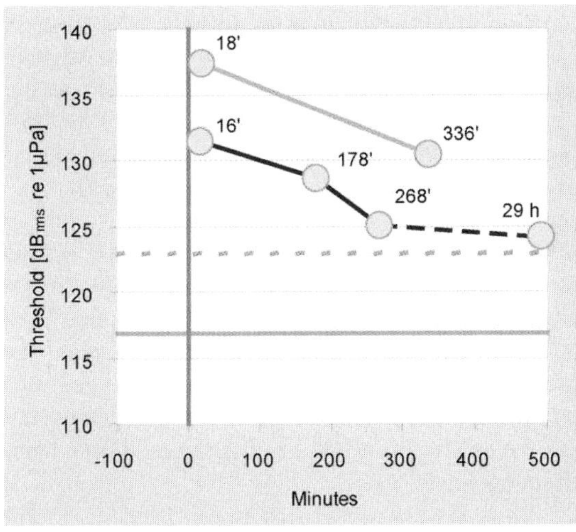

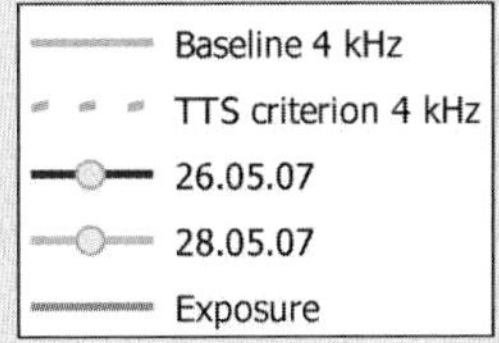

Figure 10: Recovery function of Eigil's hearing threshold at 4 kHz after sound-induced threshold shifts at this frequency. The full orange line indicates the average hearing threshold, the dashed orange line the upper variation limit of the hearing threshold and the vertical grey line the incident of the sound exposure.
The yellow circles show threshold values measured at different times (given in minutes next to the circles) after the exposures; blue and green lines indicate the two correlated sets of threshold measurements.

Its sensitivity at 4kHz had improved by 3.5dB 269 minutes post exposure, but only by another 1.4dB 29 hours post exposure. The same tendency was observed in general after TTS had clearly been reconfirmed (Fig. 10).

Conclusions / Discussion

The results of this study show, on the one hand, that the AEP method can be successfully applied for auditory studies on harbour porpoises even if the animals are unrestrained like *Eigil*, who was actively swimming and free to leave the experiments at any time.

His constant movement during the experiments, on the other hand, caused strong myogenic potentials, which were recorded along with the auditory potentials during the experiments. Even though the neuronal signals from the muscles do not occur at the same interval and repetition rates as the auditory evoked potentials, these myogenic potentials are strong enough to raise the overall neuronal noise level of the recorded potentials. At stimulus intensities close to the limit of *Eigil*'s hearing sensitivity, the auditory evoked potentials are very small because the amplitude of those potentials is a function of the perceived signal intensity, i.e. the intensity above its hearing threshold. Moreover, this function is non-linear, and only the lowest response levels are commonly included into the regression analysis, which then reveals the thresholds' values. Any masking of these lowest levels of the auditory potentials by other electrophysiological signals, such as the myogenic potentials, could obscure the real lower end of the regression line, hence leading to a 0-crossing of the regression at a higher threshold value.

The statistical analysis of the resulting EFR data may also have caused an increase in hearing thresholds. The application of a F-test to all data sets served to ensure that only AEP recordings containing an EFR at the selected modulation frequency (1.2 kHz) were included in the threshold analysis. Consequently, any AEP recordings with an EFR at or below the average neuronal noise level were excluded from the threshold analysis, potentially leading to higher threshold levels, too.

Another factor that may have influenced the hearing thresholds is the level of background noise in Kerteminde harbour. It is most likely that this broadband noise masked perception of the AEP stimuli by *Eigil*. Acoustic events, like boats passing at close distance to the research station, were avoided during the experiments by taking a time-out. Nevertheless, it was impossible to conduct the experiments at a consistently low level of background noise. As these conditions varied within each research session, and with extreme noise events excluded, one may assume that roughly the same overall noise conditions applied for all the sessions.

Despite these physical factors affecting the thresholds, the results may well reflect a hearing deficit of which *Eigil* either suffered due to a previous unmonitored exposure to intense sound or an unnoticed infection of his auditory system, or which could have developed from a long-term exposure to sounds, e.g. from the nearby harbour. However, it can be ruled out that the elevated thresholds are the result of ototoxic drugs because *Eigil* never received such treatment. An age-related hearing deficit is unlikely, too, as it usually occurs at high frequencies rather than being distributed evenly over the entire functional hearing range. *Eigil*'s hearing deficit stretches over the high and low frequencies.

As a consequence of this physiological and physical masking, the measured hearing thresholds cannot be regarded as absolute but should be defined as masked thresholds, and, accordingly, the documented threshold shifts have to be regarded as masked temporary threshold shifts (MTTS). This formalistic differentiation has no influence on discussion of the noise exposure criteria because tolerance of the auditory system is not affected by this relatively small amount of additional noise compared to the fatiguing sounds. Hence, the threshold shift limit is the same whether perception of the sounds is masked or not.

Another important result are the observed behavioural reactions of *Eigil* to the fatiguing stimuli. The fact that *Eigil* was swimming away from the incoming sound in a straight line can only be interpreted as avoidance or flight behaviour. In a free-ranging animal this reaction might have lasted over a longer period of time than observed in *Eigil*, who calmed down and was back under behavioural control of the trainers after several seconds when he was sent to subsequent hearing tests.

It also remains questionable whether or not the level of 174dB re 1μPa peak pressure or a SEL of 145dB re 1μPa^{2}s can be applied as threshold limit for behavioural reactions to impulsive sounds in harbour porpoises in general. It seems more likely that this limit varies individually and may be context specific.

So far, the only available data on behavioural reactions in harbour porpoise to sound came from observations during the construction of wind turbines at Horns Rev, Denmark (Tougaard et al. 2003), when at a distance of 15km a movement directed away from the sound source was observed in the animals, and can be derived from the BROMMAD-study (Gordon et al. 2000), when no obvious behavioural reaction was observed in free-ranging harbour porpoises in response to airgun exposures. In this context, the results of the present study constitute the first behavioural threshold in harbour porpoises that was measured under controlled acoustic conditions and can at least be used as a first indication of a threshold range.

The disturbing nature of this sound to harbour porpoises at the given intensities is underlined by the avoidance behaviour that was observed in *Eigil* prior to the exposures after the exposure level had passed his behavioural threshold for the first time. The fact that *Eigil* was trying to stay away from the monitoring hydrophone showed that he was sensitised, which was a lasting effect as he showed no signs of habituation during the remaining exposures.

TTS experiments in terrestrial animals showed that an exposure to a narrow-band signal most likely leads to auditory effects near the centre frequency of the fatiguing stimulus, while experiments involving broadband stimuli may lead to threshold shifts at a hearing frequency in the middle of the subjects functional hearing range. The results of this study indicate that this concept is in principle also applicable to marine mammals as the exposure to a narrow-band signal, such as the airgun impulse – with the bulk of its energy centred below 1kHz – led to a threshold shift at 4kHz. Nevertheless, the slight elevation of *Eigil*'s hearing threshold at 32kHz after an exposure at a level above the TTS limit indicate that the airgun impulses contain enough energy at higher frequencies to be considered moderate broad-band signals and to elicit effects in the mid frequency range of exposed subjects as well.

The measured threshold shift represents the first data of its kind for harbour porpoises. Before these results were obtained all assessments of potential effects of anthropogenic sounds on harbour porpoises had to be made based on data from other odontocete species, or even terrestrial animals. Thus, the results of this study provide the first reliable information from the species at issue, and these data can be used as a baseline to define noise exposure criteria for this species and odontocetes with acoustic and auditory characteristics comparable to the harbour porpoise. In this context, it should be noted that the resulting levels differ strongly from previous data on other species such as the bottlenose dolphin or the beluga. These two species are categorised as mid-frequency odontocetes as their sound emissions are centred in the mid frequency range of all odontocetes. They are also larger than harbour porpoises and, based on results from experiments involving blast sounds, Ketten et al. already hypothesised that the auditory effects of impulsive sounds may show a size dependency in its dose-response function. These data prove this hypothesis to be a valid concept, though a larger number of animals needs to be tested before this correlation can be clearly defined.

The limitations of these results are obvious: Only one animal has been tested so far, and it may turn out that younger animals are even more sensitive to impulsive sounds than older animals. Also, any noise exposure criterion based on these results would only be applicable to single impulse exposures. Tolerance limits for either exposures to continuous sound or repetitive exposures to impulses may be derived from these results only with strong reservations. Those levels should best be obtained from dedicated studies, and as soon as possible.

Nevertheless, these data take us a step further because they allow us to model the effect of multiple exposures to impulsive sounds on a more reliable basis than before. With the above reservations in mind, we can start modelling by using the following parameters:

Repetition rate:	Two strikes per minute within the first two minutes (as the adjustment of the pile initially takes some time), followed by 2000 strikes at 30 strikes per minute
Spreading loss:	16log r (a typical value for the North Sea; with r = distance in m)
Behaviour:	Porpoises will leave the area as soon as the ramming of foundations for a wind turbine begins
SEL level for MTTS:	164dB re $1\mu Pa^2s$
Swim speed:	6m/s (the maximum measured in a free-ranging harbour porpoise, see Lucke et al. 2000)

Several different noise exposure criteria for harbour porpoises were discussed previously in the context of windmill construction. The German Federal Environment Agency suggested limiting the exposure of marine mammals to a SEL of 160dB re $1\mu Pa^2s$ at a distance of 750m from the piling site, thereby creating a safety zone around the construction site (which would need to be monitored to ensure that no marine mammals are within this zone and hence exposed to higher levels).

Applying this stand-off distance of 750m from the sound source to the simple modelling results in a maximum allowable source level of 191dB re $1\mu Pa^2s$, a level which is reached during comparatively small-scale piling operations. All operations leading to a higher SEL may cause TTS in harbour porpoises under the given circumstances because the acoustic exposure accumulates in the hearing system of the animals and reaches the tolerance limit (MTTS) even if the animal is already kilometres away from the sound source.

This situation is drastically changed if data on the tolerance to multiple exposures from terrestrial animals are applied to this model. Based on auditory experiments involving numerous repetitions of the fatiguing stimulus under controlled conditions, it has been shown that the TTS limit drops by 3dB per ten-fold increase in the number of exposures. If we take this into account for the model, a maximum source level of 182dB re 1μPa^2s results.

Any alteration in the parameters of the model would of course change these maximum source levels. The swim speed of the porpoises e.g. represents the maximum value measured for an adult harbour porpoise. It is unknown for how long harbour porpoises can maintain this speed and it is questionable whether or not juveniles would be capable of swimming at this speed at all.

A potential mitigation method which has been suggested for piling operations is the so-called 'soft start', a slow ramp-up of the source level over an extended period of time. Such a soft start with a reduced repetition rate of the piling impulses (e.g. 2 per minute) over the first 15 minutes and an initial SEL of 150dB re 1μPa^2s would increase the maximum allowable source level to 188dB re 1μPa^2s without causing any TTS in harbour porpoises.

Even though soft starts are a default mitigation method in seismic operations, their effectiveness has never been proven. It is still unknown whether or not animals leave the area because of the increase in received level, i.e. whether or not they can detect this gradient and react accordingly.

Other mitigation methods which could be applied during piling operations exceeding the noise exposure criteria are the use of so-called pingers for deterrence of harbour porpoises or air-bubble-curtains around the sound source. While pingers emit sounds that have been proven to drive harbour porpoises away from their vicinity, air-bubble-curtains can effectively reduce the amount of sound energy radiating from the sound source. The same principle applies for the application of bubble-foam layers which are wrapped around the piles. Other mitigation measures include engineering solutions such as prolongation of the ramming impulse and the temporal and spatial closure of areas for ramming activities.

Acknowledgments:

These experiments would have been impossible without the help and support of numerous people. We would like to acknowledge CGG Veritas, France, for providing the airgun and Alain Regnault for his patient support with this device. Wolfgang Voigt, Research and Technology Center Westcoast in Buesum, provided valuable support in this respect, too. The staff of the Fjord- & Bælt was exceptionally helpful and patient over the whole study period. Kristian Beedholm and Lee Miller from the University of Southern Denmark, Odense, generously provided ongoing logistic and intellectual support. We would also like to thank the source boat team, Jacob Rye Hansen, Cecilia Vanman, Mario Acquarone, Heiko Charwat and all volunteers. Important parts of the equipment used in these experiments were provided by the Bundeswehr Technical Center for Ships and Naval Weapons (WTD 71) in Eckernförde and the Plön measurement site, as well as by the GKSS Research Centre in Geesthacht. Their support is greatly appreciated. The experiments were conducted under permit from the Danish Forest and Nature Agency, Denmark.

References

Andersen S (1970). Auditory Sensitivity of the Harbour Porpoise *Phocoena phocoena*.. In: Investigations in Cetacea, Vol.2:255-259

Finneran JJ, Schlundt CE, Dear R, Carder DA, Ridgway SH (2002). Temporary shift in masked hearing thresholds in odontocetes after exposure to single underwater impulses from a seismic watergun. J. Acoust. Soc. Am., Vol. 111:2929-2940.

Gordon J, Freeman S, Chappell O, Pierpoint C, Lewis T, MacDonald D (2000). Investigations of the effects of seismic airguns on harbour porpoises: Experimental exposures to a small source in inshore waters. In: Thompson D (Ed.): Behavioural and Physiological Responses of Marine Mammals to Acoustic Disturbance (BROMMAD). Final Scientific and Technical Report, University of St. Andrews, St. Andrews, UK

Kastelein RA, Bunskoek P, Hagedoorn M, Au WWL (2002). Audiogram of a harbour porpoise (*Phocoena phocoena*) measured with narrow-band frequency-modulated signals. J. Acoust. Soc. Am., Vol. 112(1):334-344

Lucke K, Wilson R, Teilmann J, Zankl S, Adelung D, Siebert U (2000). Advances in the elucidation of cetacean behaviour: A case study on the harbour porpoises (*Phocoena phocoena*). In: Teilmann J (2000): The behaviour and sensory abilities of harbour porpoises (*Phocoena phocoena*) in relation to bycatch in gillnet fishery. Ph.D. thesis, University of Southern Denmark, pp.87-105

Nedwell J; Turnpenny AWH, Longworthy J, Edwards B (2003). Measurements of underwater noise during piling at the Red Funnel Terminal, Southampton, and observations of its effect on caged fish. Subacoustech Ltd. 558 R 0207. 2003.

Ødegaard & Danneskiold-Samsøe A/S (2000). Offshore Wind-Turbine Construction. Offshore Pile-Driving Underwater and Above-Water Noise Measurements and Analysis. McKenzie Maxon, Christopher and Winther Nielsen, Ole. 00.877.

Popov VV & Supin AYa (1990). Electrophysiological studies of hearing in some cetaceans and manatee. In: Thomas JA & Kastelein RA (Eds.): Sensory Abilities of Cetaceans: Laboratory and Field Evidence. Plenum Press, New York, U.S.A. pp.405-415

Robinson S, Lepper PA, Ablitt J (2007). The Measurement of the Underwater Radiated Noise from Marine Piling including Characterisation of a "Soft Start" Period". Oceans 07 IEEE Aberdeen Conference Proceedings, Aberdeen, Scotland.

Southall BL, Bowles AE, Ellison WT, Finneran JJ, Gentry RL, Greene CR Jr, Kastak D, Ketten DR, Miller JH, Nachtigall PE, Richardson WJ, Thomas JA , Tyack PL (in press). Marine mammal noise exposure criteria. Aquat. Mamm.

Tougaard J, Carstensen J, Henriksen OD, Skov H, Teilmann J (2003). Short-term effects of the construction of wind turbines on harbour porpoises at Horns Reef. Technical report to TechWise A/S. HME/362-02662. Hedeselskabet, Roskilde.

Excursus 4: Harbour seal

Katrin Wollny-Goerke, Ursula Siebert

The **harbour seal** or **common seal** (*Phoca vitulina vitulina*) belongs to the suborder *Pinnipedia* (pinnipeds) and superfamily *Phocoida* (seals). Different subspecies are widely distributed throughout the Atlantic and Pacific coastal regions of the northern hemisphere. Harbour seals are the most abundant seals in German waters, living in the North Sea in high numbers while being rarely seen in the German Baltic Sea. The stocks in the North Sea have thankfully recovered after the Harbour seal epidemics in1988 and in 2002. Trilaterally coordinated, aerial surveys of the seals on their haul-out sites in the Wadden Sea and counts on Helgoland are made about five times a year.

Harbour seals are the best-known representative of the phocid seals in Germany. They grow to a length of 1.90m and can weigh up to a maximum of 120kg. They reach sexual maturity at the age of 3 to 4 years. Mating takes place in summer, immediately after weaning the pup. Implantation is delayed for 1.5 to 3 months, and then 9 months later the female normally gives birth to one pup in early summer. In the Wadden Sea, the birth period begins at the middle / end of May and lasts until the beginning of July, with a peak of 95% of births in June. The pups are born on tidal flats or undisturbed sandy beaches. They can swim immediately after birth and follow their mothers. The mother recognizes her pup by its vocalisations. Birth weight is about 8 to 10kg. The lactation period only lasts about four weeks.
Haul-out sites are particularly used during the birth and lactation period, and the annual moult. The latter takes place during July and August. Seals also use the sandbanks for resting during other times of the year, though to a lesser extent.

Harbour seals are opportunistic feeders that mainly prey on different flatfish species and other demersal fish, depending on the prey availability. Young seals also feed on shrimp. Harbour seals find prey primarily with the help of their vibrissae, which enable them to locate fish or water turbulences caused by fish movements. Seals travel long distances into the North Sea for feeding. Sometimes they remain in an area for several days; then they swim quickly between these areas, without changing direction (see chapter 5).

Harbour seals, like other pinnipeds, produce sounds in the air and underwater for communication. Vocalisation plays an important role in maintaining contact between mother and pup, between mating partners and for territorial behaviour.

The Habitats Directive is the most important international legal instrument for the strict conservation of seal species occurring in the North and Baltic Seas. Furthermore, harbour seals are under special protection of the Bonn Convention (CMS) with its Wadden Sea Seal Agreement and the associated Seal Management Plan.

5 Determination of space and depth utilization of the Wadden Sea & adjacent offshore areas by harbour seals

Gabriele Müller, Dieter Adelung, Nikolai Liebsch

Zusammenfassung

Seehunde sind eine der größten Attraktionen im Wattenmeer. Der geplante Bau von Offshore Windkraftanlagen in der deutschen AWZ (Ausschließliche Wirtschaftszone) könnte eine Beeinflussung der Seehunde darstellen. Mittels eines Satelliten-gestützten Fahrtenschreibersystems wurde die räumliche Nutzung des Wattenmeeres und der angrenzenden Offshore-Gebiete durch die Seehunde untersucht. Neben der Ermittlung der Schwimmrouten gab die spezifische Analyse des Tauchverhaltens Auskunft über die Nutzung des Habitats in allen drei Dimensionen. Die Tiefenprofile von einzelnen Beutezügen spiegeln die Veränderungen in der Tiefe des Meeresbodens wider, da die Seehunde in fast allen Tauchgängen bis zum Boden tauchen. Seehunde aus dem Wattenmeer bevorzugen Tiefen zwischen 12 und 25 m, während Seehunde von Helgoland ein weitaus größeres Tiefenspektrum (bis 60 m) nutzen.

Die Ergebnisse zeigen, dass die Seehunde den größten Teil ihrer Zeit in Offshore-Gebieten zwischen 30 und 60 km von Land verbringen, aber auch Gebiete bis zu 150 km entfernt zur Nahrungssuche nutzen. Seehunde von Helgoland nutzen primär das Gebiet bis zu einem Radius von 25 km um die Insel. Somit zeigen die von den Seehunden genutzten Gebiete eine Überlappung mit einigen der geplanten Windkraftanlagen. Angesichts der geringen Überlappung dürften jedoch genügend Ausweichmöglichkeiten für die Nahrungssuche zur Verfügung stehen, wobei dies eventuell zu einer erhöhten Nahrungskonkurrenz mit anderen Seehunden, Kegelrobben und Schweinswalen in den betroffenen Gebieten führen kann. Das Wattenmeer selbst dient vorrangig als Ruhezone.

Abstract

Harbour seals are one of the main attractions in the Wadden Sea area. The planned construction of large offshore wind farms in the German Exclusive Economic Zone (EEZ) has potential to affect these seals. Using a satellite-supported dead-reckoning system, the space and depth utilization of the Wadden Sea and adjacent offshore areas by harbour seals was investigated. Beside the analysis of the foraging routes, a quantitative analysis of the diving behaviour of harbour seals provides information on the utilization of the three-dimensional habitat that the seals live in. The depth profile of individual foraging trips reflects the depth of the sea bed as seals usually dive to the bottom. Seals from the Wadden Sea prefer depths between 12 and 25m, while seals from Helgoland use a much wider depth range (up to 60m).

The results show that the seals spend most of their time at sea in areas between 30 and 60km from the Wadden Sea, but also use areas of up to 150km from shore. These regions are their main foraging areas. The Wadden Sea serves primarily as resting zone and for pupping and moulting. Seals from Helgoland spend most of their time within 25km of the island. The tagged seals' distribution therefore overlaps with the easternmost planned and / or approved offshore wind farms. Based on the small overlap, there should be enough alternative foraging areas available though increased competition for food with other harbour seals, harbour porpoises, and grey seals in the affected areas may result.

Introduction

The harbour seal (*Phoca vitulina*) is a common sight in the Wadden Sea and a major tourist attraction. It is a top predator, feeding mainly on bottom-associated fish. High hunting pressure has reduced the population size to a few thousand animals, compared to 36,000 at the beginning of the 19th century (Reijnders 1992). Since hunting was finally prohibited in all areas of the Wadden Sea in 1976, the population size has increased almost continuously. In 1988 and 2002, the population was severely struck by a phocine distemper virus (PDV) epidemic. Between 40 and 60% of all animals died during these outbreaks. The increase in numbers of counted animals on the sand banks (by simultaneous aerial surveys) to 15,426 in 2006 (CWSS 2006) can be linked to birth rates of up to 18.5% (Hasselmeier 2006). Accounting for the foraging – and thus not visible – seals, the total number is presumably near 20,000 animals.

Harbour seals were commonly associated only with the Wadden Sea because they could almost always be observed there (Mohr 1952). Human activities outside the Wadden Sea area were assumed not to affect the seals. Recent studies, however, have shown that harbour seals spend a significant amount of time outside the Wadden Sea in offshore areas (Orthmann 2000, Dietz et al. 2003, Tougaard et al. 2003, Adelung et al. 2004, Liebsch et al. 2006, Tougaard et al. 2006). The German Bight is an extensively used marine area, with types of use ranging from oil and gas production, commercial and recreational ship traffic to fisheries and gravel extraction and the concomitant input of waste and pollutants (NPA & UBA 1998). The construction of large offshore wind farms will be a new type of use for this habitat and the consequences are difficult to assess. One of the main reasons is the lack of data concerning habitat use and movement patterns of harbour seals but also their hearing abilities (see chapter 6). In the MINOS project a method was developed to obtain this type of data (Adelung et al. 2004) and the MINOS+ project aimed at enlarging this data base as well as obtaining information during the construction phase of approved offshore wind farms.

The satellite-supported dead-reckoning system used in this study permits the reconstruction of harbour seal swimming routes. In addition, it provides a detailed record of the diving behaviour, which holds information on the animal's use of the third dimension. From this type of data conclusions can be drawn regarding e.g. the depths at which harbour seals find their prey or the time spent at the surface.

Methods

The study of behaviour and space utilization of pinnipeds at sea has been promoted significantly by the development of telemetry technology (Kooyman 2004). Satellite telemetry primarily allows the determination of animal location at sea, whereas multi-channel data loggers collect information on the diving behaviour of the animal. To study the space utilization and behaviour of harbour seals from the Wadden Sea, a satellite-supported dead-reckoning system was developed with Driesen & Kern GmbH (Germany) (Fig. 1). This system consists of a dead-reckoner, a satellite transmitter, and, in autumn and winter, of an electronically timed release mechanism. All instruments were packed into a housing that served as a floating device and protection.

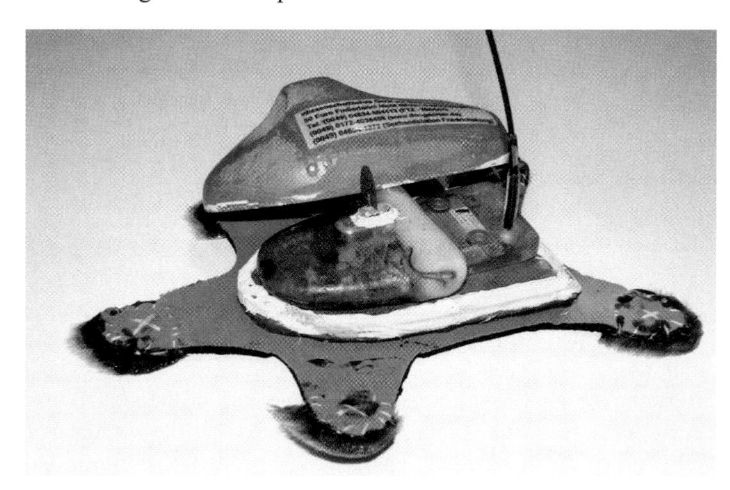

Figure 1: Satellite-supported dead-reckoning system with floating device (red), neoprene base (blue), dead reckoner and satellite tag (with antenna). Note the fur attached to the tips of the neoprene base due to the release during moult.

The dead-reckoner is a multi-channel data logger, sampling dive depth, speed, heading, body orientation, pitch, roll, temperature, and light. From this information, the three-dimensional dive path of the seal can be calculated (Mitani et al. 2003). In combination with the satellite tag, that provides information on seal haul-out locations and thus start and end points of foraging trips, the routes of individual foraging trips can be reconstructed. With a memory capacity of 32MB and a sampling interval of 5 seconds, the dead-reckoners are able to collect continuous information for up to 94 days, allowing the monitoring of individual animals for a longer period.

Satellite transmitters were obtained from Wildlife Computers (Redmond, USA): SPOT2 tags used in the early phase of the study were later replaced by the smaller SPOT3 and SPOT4 tags. A salt water switch determined the on/off status of the tag to limit transmissions to periods when the tag is dry. Due to the attachment site of the device (i.e. the lower back), the satellite tag was generally only dry when the seal was on land. Thus, the majority of locations were obtained from haul-out locations. The accuracy of usable locations was generally between 100 and 1000m and thus sufficient to identify haul-out sites and locate devices after their release from the animal.

The housing consisted of a mixture of resin and microscopic glass beads and was coloured red to aid in recovery. The size was minimised as much as possible without loosing the slightly positive buoyancy necessary for floatation after the release from the animal. Systems deployed in spring were released by the annual moult that occurs in July and August. In autumn and winter the housing was inserted into a neoprene pocket to which it was anchored by a nylon cord. The electronic release mechanism was programmed to activate a current between an anode and a small stainless steel rod (acting as the cathode – both connected to a battery) holding the nylon cord. This current caused the corrosion of the rod, thereby releasing the housing. The neoprene pockets (autumn/winter) and bases (spring) were attached to the animal's fur on the lower back with a two-component epoxy glue (Fig. 2).

Figure 2: Harbour seal on Helgoland with tracking system. The floating device (red) and the neoprene base (blue) will fall off during the annual moult.

By only gluing small 2 x 2cm patches to the fur, the freedom of movement of the animal was not significantly affected. After release from the animal, the devices float at the surface and usually wash ashore in the German or Danish Wadden Sea due to the prevailing currents and winds. These areas are highly frequented by people walking along the beaches, so that most devices are found and returned.

Harbour seals were captured at three different locations: on the sand bank Lorenzenplate (54°38'N / 8°53'E), on Rømø (55°21'N / 8°50'E, in the Danish part of the Wadden Sea), and on the offshore island Helgoland (54°19'N / 7°92'E). On the Lorenzenplate and on Rømø, seals were captured with a 100 m long seine net from the sea side as described in detail by Jeffries et al. (1993). Because seals on Helgoland are accustomed to tourists, they could be approached on foot and caught by hand.

Results

Between June 2004 and December 2006, a total of 73 seals were equipped and 55 devices could be recovered, resulting in a recovery rate of 75% (Table 1).

Table 1: Number of harbour seals equipped at three different locations, with the number of recovered devices in parentheses.

		Spring	Autumn	Winter
Lorenzenplate	females	6 (4)	3 (2)	3 (3)
	males	10 (9)	5 (4)	3 (1)
Rømø	females	-	-	1 (1)
	males	15 (9)	6 (5)	3 (1)
Helgoland	females	3 (3)	-	-
	males	7 (6)	8 (7)	-

35 of the recovered devices contained usable data (i.e. dive data from at least one complete foraging trip), resulting in 332 foraging trips and 1353 seal days. A total of 279 routes could be reconstructed for 30 seals (24 males and 6 females).

Haul-out locations

The Wadden Sea is used extensively by harbour seals as resting area between foraging trips. Many haul-out sites are located on sand banks in the outer Wadden Sea area closer to offshore foraging areas (see below). Individual animals may use haul-out sites located in the more sheltered inner areas of the Wadden Sea, e.g. a pregnant female that moved from the outer to the inner Wadden Sea area during the pupping and lactation phase.

The number of haul-out sites used by individual animals (during a period of up to 94 days) varies between one and eight. Often these sites are in close proximity to each other. Less frequently seals use sites that are far apart. An example for the latter is a seal equipped on Helgoland in autumn 2006. This seal subsequently used haul-out sites in the Wadden Sea east of Helgoland and then moved westward to haul-out sites on Texel in the Netherlands.

Utilization of offshore areas

Harbour seals alternate periods on land with time spent at sea searching for prey. A foraging trip is characterised by a movement into deeper water away from the haul-out site and a more or less continuous diving activity. The duration of these foraging trips depends on whether the seals start in the Wadden Sea or from Helgoland as the foraging strategies differ markedly between these sites. Foraging trips originating in the Wadden Sea have a mean duration of 4.6 days but may last between 0.5 and 13.5 days. In contrast, almost all trips starting on Helgoland last less than 24 hours and only very rarely exceed 2 days. The routes of seals equipped at the three different locations are presented in Fig. 3.

When departing from haul-out sites in the Wadden Sea, harbour seals move westward into deeper water. The seals alternate periods of rather straight movement with periods of more or less intense meandering. This pattern is very obvious in seals from Rømø (Fig. 3B), where foraging trips start with a directed movement out of the Wadden Sea. After reaching deeper water the animals start meandering, which is characteristic of foraging activity. The meandering can be interspersed with shorter or longer periods of directed movement when animals move between prey patches. This is particularly the case with seals from the Lorenzenplate (Fig. 3A). These seals also often use Helgoland as haul-out site, as indicated by the high number of routes around the island.

The short duration of trips from Helgoland implies that the seals do not move very far from the island. The vast majority of routes lie within 25km of the island, with a maximum of 40km (Fig. 3C).

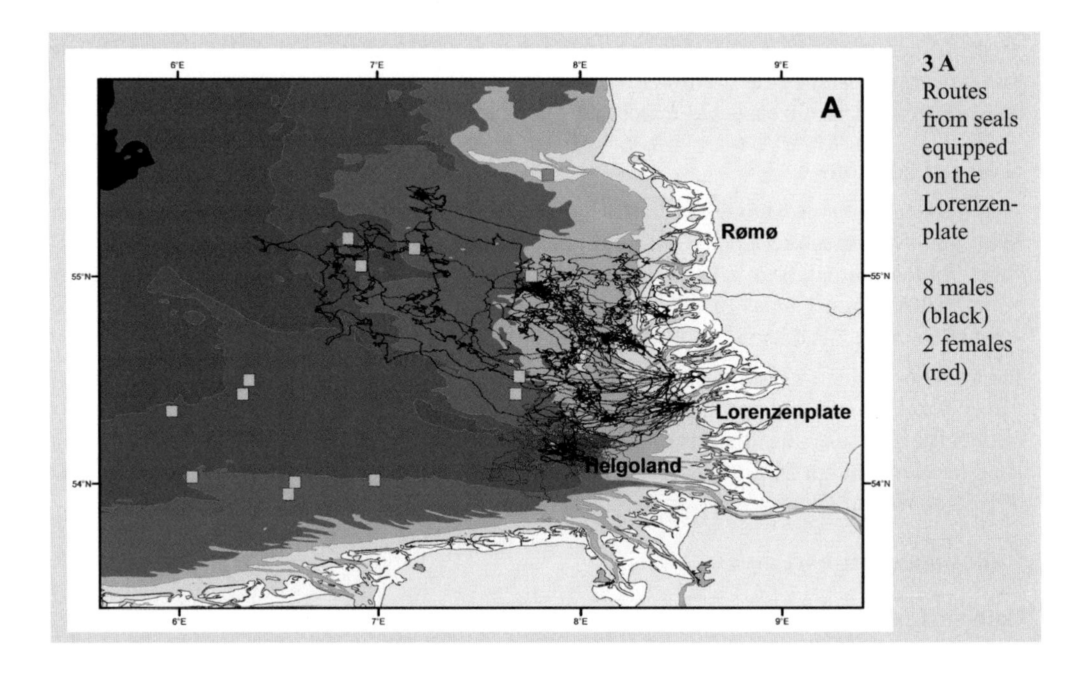

3 A
Routes from seals equipped on the Lorenzen-plate

8 males (black)
2 females (red)

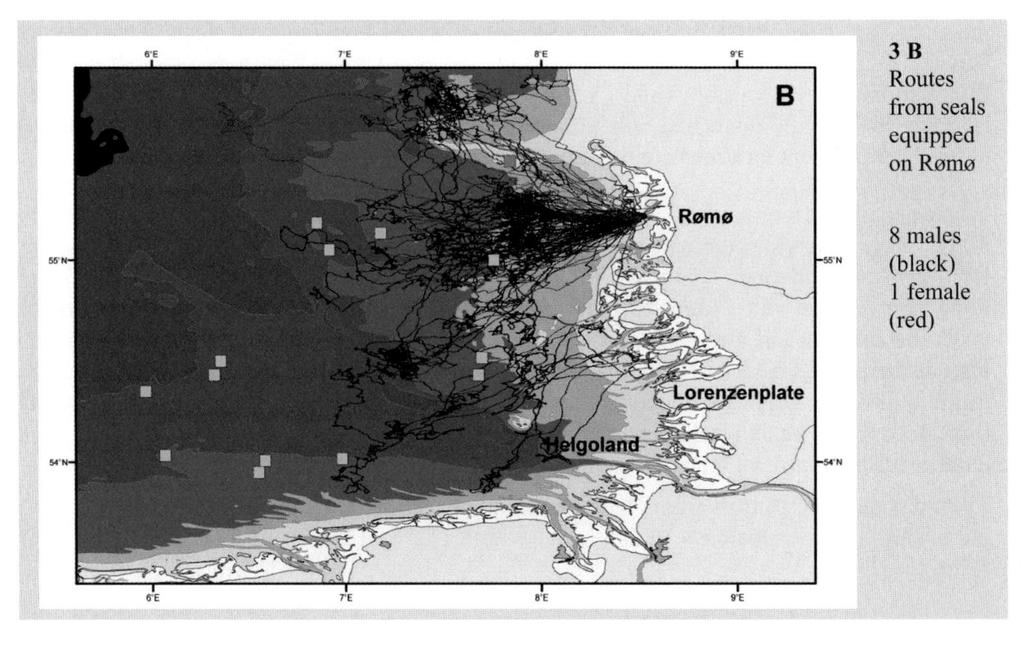

3 B
Routes from seals equipped on Rømø

8 males (black)
1 female (red)

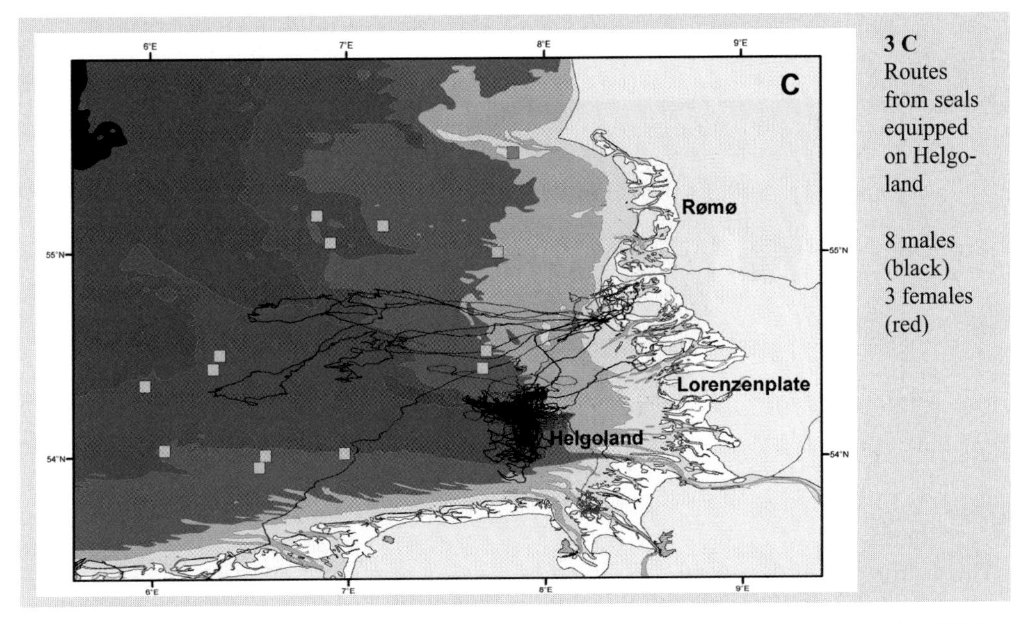

Figure 3: Routes of foraging trips made by harbour seals (between June 2004 and December 2006). Orange squares: approved and planned offshore wind farms in the German Exclusive Economic Zone. Green square: operational Danish wind farm Horns Reef.

Only seals that have moved from Helgoland to the Wadden Sea make trips to more distant areas. The range of seals from the Wadden Sea is markedly larger, as they generally spend most of their time between 30 and 60km from their haul-out sites. They may, however, move up to 150km from land.

When combining the routes from all locations, it becomes apparent that harbour seals mainly use the area between 30 and 60km offshore as well as the area (up to 25km) around Helgoland. These areas overlap only with the easternmost of the planned and/or approved wind farms. One animal from Rømø moved repeatedly in a north-westerly direction and passed the offshore wind farm Horns Reef on the way to its foraging grounds.

Diving behaviour

A total of 332 complete foraging trips were recorded and the 308.475 dives associated with these trips were used in the analysis.

Individual trips show a general pattern with depth increasing after departure from the haul-out site and usually reaching maximum values during the central part of the trip (Fig. 4). Thus, the depth profile of a foraging trip reflects the depth of the sea bed with the deepest dives furthest away from the haul-out site (see below). A few rather deep dives at the start or end of the trip may be associated with swimming through deeper tidal creeks when leaving the Wadden Sea.

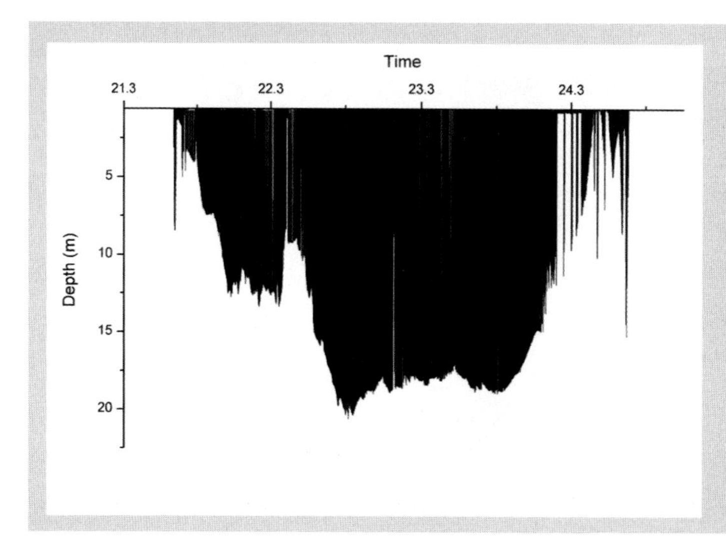

Figure 4: Depth profile of a 3-day foraging trip made by a seal from Rømø, DK, reaching maximum depths of approximately 20m. Individual dives appear as thin lines.

The distribution of dive depth from all trips shows a significant difference in depth usage between seals from the Wadden Sea and those from Helgoland (Fig. 5). Seals from the Wadden Sea primarily use depths between 12 and 25m and only rarely dive deeper than 30m. In contrast, seals from Helgoland use a much broader depth range and reach almost 60m, even though the range from 30 to 60m is not used as extensively as the shallower depths.

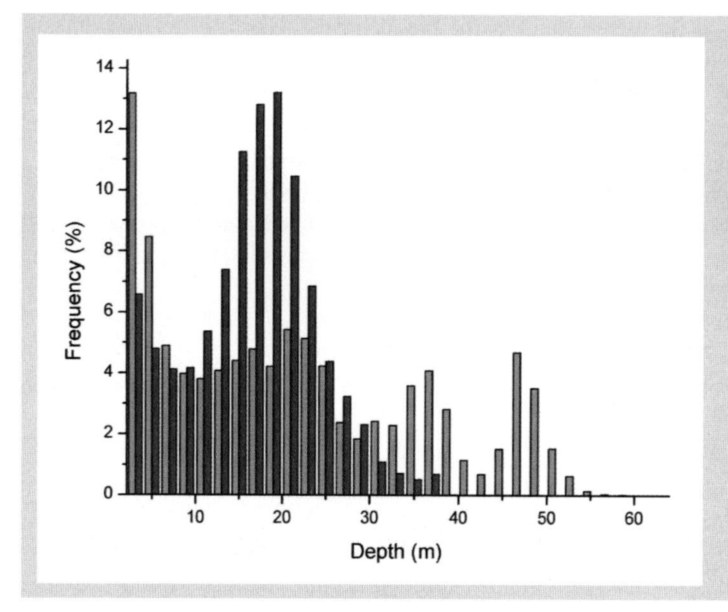

Figure 5: Comparison of distribution of dive depth of 308.475 dives associated with foraging trips from seals from the Wadden Sea (red) and from Helgoland (grey).

The combination of location and depth allows the depiction of a foraging trip in three dimensions (Fig. 6).

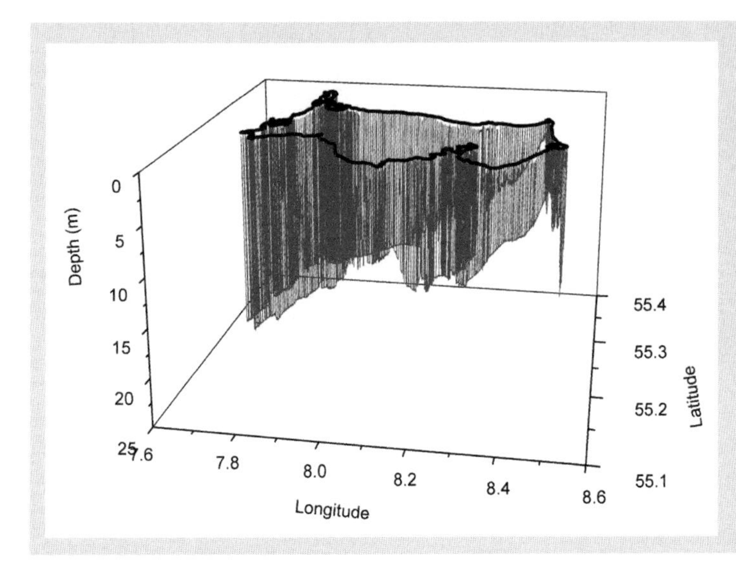

Figure 6: 3D-Route made by a seal from Rømø, DK (same trip as in Fig. 3 above). The start and end point of the trip is located to the right in the figure. The black line shows the foraging trip in two dimensions.

It becomes apparent that the deepest dives occur farthest from the haul-out site and that the change in depth (compare the depth profile of this trip in Fig. 4) concurs with movements away from or towards the haul-out location.

A closer look at individual dives reveals different dive profiles. Most of the dives made by harbour seals during a foraging trip are U-shaped dives with three distinguishable phases: a descent phase where the seal descends from the surface to the bottom, a bottom phase where the seal follows the sea bed, and an ascent phase where the seal returns to the surface. This pattern resembles the letter "U" and therefore these dives are called U-shaped dives. The second dive type that is seen regularly (though much less frequently than U-shaped dives) is the V-shaped dive, which consists only of a descent and an ascent phase.

The two-dimensional depth profiles of all U-shaped dives are more or less similar. However, when looking at additional parameters such as pitch and roll, a range of different "sub-types" emerge (Fig. 7A, B). When seals leave their haul-out site or move between prey patches, they frequently use so-called "travel dives" (Fig. 7A). In these dives the animal descends, follows the sea bed, and ascends with a more or less constant heading and thus covers a large distance between the start and end point of the dive. When a prey patch is encountered, the seal may switch to "foraging dives" (Fig. 7B) where the descent and ascent phases are similar to those in travel dives but the bottom phase is characterized by a high variability in movement and heading. In addition, the distance covered during such a dive is usually much less than in a travel dive. A third type of U-shaped dive is the "resting dive." Here, the seal sinks to the bottom and lies more or less motionless on the sea floor before ascending. V-shaped dives are variable in depth (particularly when compared to preceding or subsequent U-shaped dives). They have been associated with exploratory behaviour, but the ultimate function still remains to be determined.

The average dive duration of harbour seals lies between 3 and 5 minutes, with resting dives lasting up to about 15 minutes. While seals are able to dive for about half an hour, they rarely

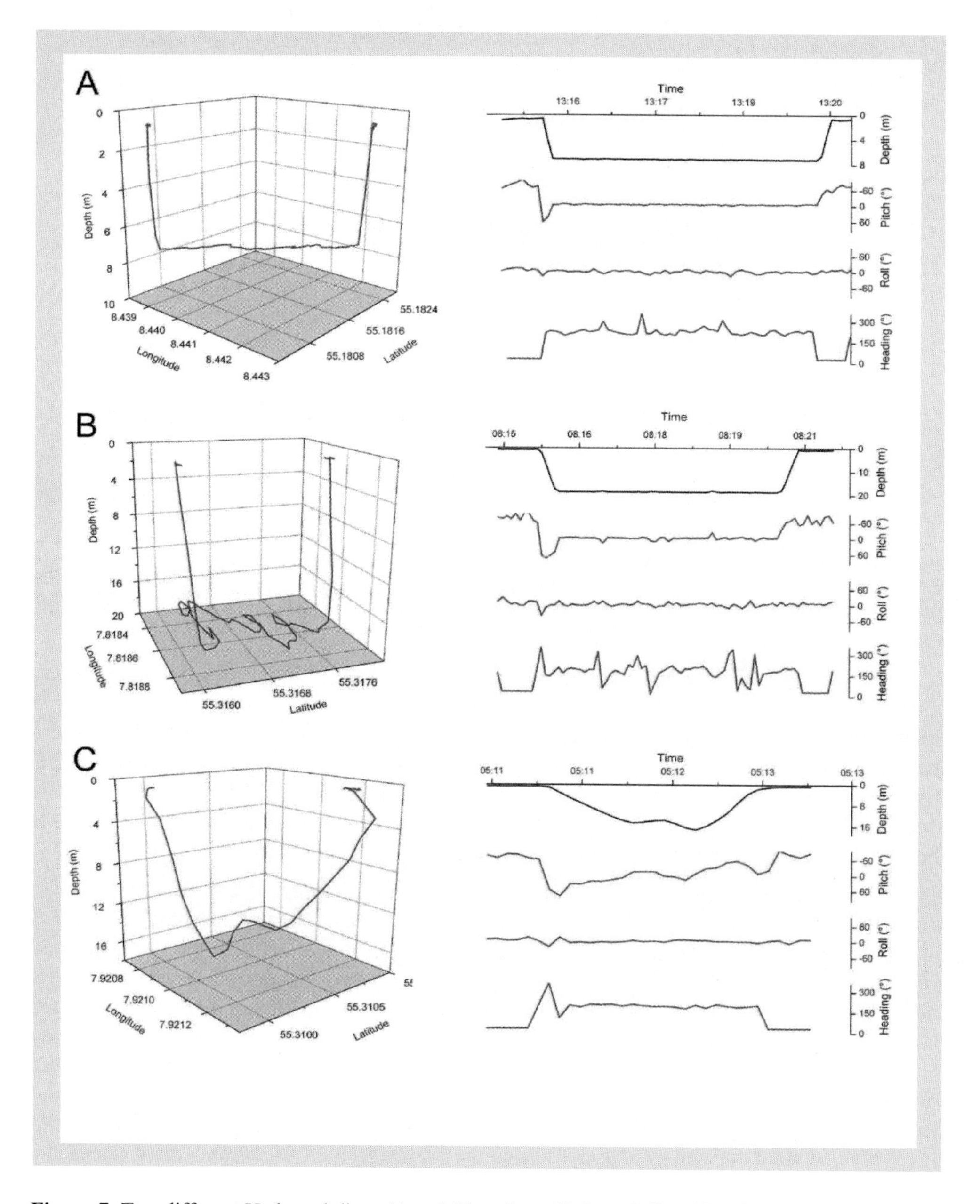

Figure 7: Two different U-shaped dives (A and B) and one V-shaped dive (C) with the according three-dimensional dive profiles (left) and corresponding two-dimensional presentation of depth, pitch, roll, and heading against time (right). Note the similarity of the two-dimensional depth profile of the two U-shaped dives.

exceed 6 minutes. Between dives the seals spend approximately half a minute at the surface to replenish their oxygen stores before beginning the next dive. On longer foraging trips, extended surface intervals of up to one or two hours may occur where the animal rests at the surface in a vertical position with only the head above the water.

Discussion

The available data is a good basis for evaluating potential overlap between harbour seal distribution and offshore wind farms. The number of seals that could be equipped and tracked has increased dramatically compared to the first part of the project, and the amount of information gained from individual animals has increased mainly due to the larger memory capacity in the dead-reckoners.

The Wadden Sea area is an important habitat for resting, reproductive activities such as pupping and lactating, as well as for the moulting process during which seals need a large amount of sun light. However, this study confirmed that the seals spend the majority of their time at sea outside the Wadden Sea and any activity in these offshore areas has the potential to affect these animals.

The current data show that harbour seals roam over large offshore areas and also utilise areas for which wind farms have been approved and/or planned. The distinct movement patterns of seals observed in the MINOS project, where foraging trips were divided into three separate phases based on the "straightness" of their movements (Adelung et al. 2004), was also observed in this study (primarily in the seals from Rømø), but more frequently a modified pattern occurred. It appears that the majority of seals use a more variable movement pattern and roam around a much larger area than previously thought. Here, the movements during the middle phase were not as consistent, with periods of rather straight movements alternating with periods of extensive meandering, as animals move from one prey patch to the next.

A trend of movements further offshore was observed during the autumn and winter months, but since the amount of data for this period is markedly lower than for the spring season, no definite statements can be made in this respect. In addition, the low number of females in this study precludes any comparison of movements between the two sexes. One of the farthest trips was, however, made by a female, suggesting that (in combination with the lack of significant dimorphism) they are physiologically able to exploit the same areas as males.

The available information on the diving behaviour of harbour seals provides a detailed image of the seals' activity at sea. The depth of the dives concurs with the water depth along the reconstructed routes. The seals from the Wadden Sea move out into deeper water and spend most of their time in areas with depths between 12 and 25m, which correspond well with the 30 to 60km range shown by the routes. Seals from Helgoland have direct access to deeper water and their depth utilization also reflects this larger depth range.

Diving activity is almost continuous and only interrupted by short periods of resting dives or inactivity at the surface. The use of comparatively short dives (compared to possible maximum dive durations) and subsequent short surface intervals allows the seal to spend a large fraction of its time below the surface. Due to the relatively shallow depth of the German Bight, they can therefore maximize their time at the bottom where prey can be encountered.

Long dives would require an increase in surface interval duration, resulting in a less efficient diving strategy.

The majority of dives are U-shaped, a pattern that is frequently observed in diving seals (Schreer et al. 2001, Liebsch 2006). This indicates that seals mainly take prey associated with the sea bed such as flatfish. This corresponds well with available information on adult harbour seal stomach contents (Behrends 1985, Nørgaard 1996, Tollit et al. 1998). The V-shaped dives usually do not reach the sea bed, and have been thought to represent exploratory behaviour (Adelung et al. 2004). An alternative interpretation could be the encounter of prey in the water column, but further analysis is necessary to support this hypothesis.

Interpretation of dive function was mostly based on the two-dimensional depth profile (time over depth). As most of the dives made by harbour seals are U-shaped, it is unlikely that all these dives have the same function. Additional parameters such as dive angle (pitch) and movements to the side (roll) have shown that U-shaped dives represent a range of different activities such as travelling, foraging, and resting. It is therefore important to include such information in the analysis of the diving behaviour.

The fact that harbour seals mainly forage at or close to the sea bed may be important in predicting possible effects of offshore wind farms on the seals. The area used by harbour seals mainly consists of soft bottom substrate, which does not support a very rich benthic community. The construction of wind farms introduces a hard substrate, i.e. the anchoring system of the wind mills, which has been shown to attract a larger number of fish (Leonhard & Petersen 2006). Together with the ban of fisheries, wind farm areas could thus become more attractive foraging areas for seals and other top predators. However, other factors such as the noise from the wind mills themselves as well as from the ship traffic to and from the wind park area (and the associated disturbance) could offset the advantages of the introduced hard substrate. Until wind farms will have been constructed and further data on seal behaviour collected, no definite conclusions can be drawn on the effects they will have on the seals.

The existing wind farm Horns Reef in Denmark (with 80 turbines the largest offshore wind farm worldwide) does not seem to be directly avoided by the seals. However, only one animal moved into the direction of this wind farm from Rømø. This may have been caused by the generally more westward movements of the seals when leaving the Wadden Sea. If Danish seals generally use this movement pattern, then seals from a more northerly haul-out site, Skallingen, would be more likely to be heading towards the wind farm, and any avoidance reactions could potentially be observed. Tougaard et al. (2006) studied the distribution and movements of harbour seals close to Horns Reef and showed that this area was visited by seals before construction but was not of any special interest to them. An avoidance reaction was observed during the construction phase, most likely because of the noise associated with the ramming of the monopiles. During operation of the wind farm, seals were occasionally seen and satellite signals from tagged seals were also obtained from inside the wind farm area, suggesting that seals did not avoid this area any more. However, due to the comparably low accuracy of the satellite locations, the results should be interpreted with caution. In general, the data do not suggest that this wind farm area is an important foraging area for harbour seals.

In 2003 a second and smaller Danish offshore wind farm was built near Nysted in the Fehmarn Belt (Baltic Sea). This wind farm is located about 10km from shore where the water depth is between 5 and 10m. Dietz et al. (2003) studied harbour and grey seals at a nearby seal sanctuary before, during, and after the construction phase and found that only the construction phase (i.e. the noise) caused some disturbance on the haul-out site. Disturbance of haul-out behaviour by noise is not expected for seals from the Wadden Sea due to the large distance between haul-out sites and wind farm areas.

Due to the fact that the beginning of the construction of the approved offshore wind farms in the German EEZ has been postponed for the time being, no information could be obtained on how the seals react to the increased noise and shipping traffic that would accompany this phase. However, the data do show that the areas in which wind farms are approved and planned do overlap with harbour seal foraging areas. This is particularly true for the wind farms in shallow offshore areas within 60km from shore. Such an overlap may also occur off the coast of Eastern Frisia where approved and planned wind farms are much closer to the coast.

Compared to harbour seals, grey seals (*Halichoerus grypus*) range over much larger areas (Dietz et al. 2003). The number of these seals and thus their importance is continuously increasing but, unfortunately, no data are available on their movements and diving behaviour in the German Bight.

To summarise, the current results show that harbour seals use extensive areas for foraging outside the Wadden Sea, which only partially overlap with the sites where wind farms will be built. Therefore, they should be able to temporarily switch to alternative foraging areas during the construction phase of the wind farms when the main interference can be expected. Considering the recent population size and the amount of overlap, no eminent danger is expected for the harbour seals at this stage.
An expansion of the planned wind farms and/or an increase in population size of harbour seals, grey seals or harbour porpoises (*Phocoena phocoena*), however, may lead to increased competition for food resources. As there are no data available on intra- and interspecific competition in this area, a future reassessment of the situation which includes these parameters – should population sizes increase – as well as an evaluation of possible influences of further expansions of the wind farms would be beneficial.

Acknowledgements

We would like to thank the following people for their advice and help, particularly during the field work:
- Marine mammal research team of the RTC - Büsum headed by Dr. Ursula Siebert
- Team of the Seal Centre Friedrichskoog headed by Tanja Rosenberger
- Marine mammal research team of the GKSS – headed by Prof. Dr. Andreas Prange
- Svend Tougaard and Thyge Jensen, Fiskeri- og Søfartsmuseet, Esbjerg, DK
- Jakob Tougaard and Jonas Teilmann, National Environmental Research Institute, Roskilde, DK
- Dr. Thomas Borchardt, Office of the Wadden Sea National Park in Schleswig-Holstein, Tönning
- Karl-Heinz Hildebrandt, National Park Service, Tönning
- Dr. Gerd Meurs, Multimar Wattforum, Tönning
- Mandy Kierspel, Leibniz-Institute for Marine Sciences, Kiel

- Dr. Kai Abt, Kiel
- Veterinarians Dr. Jörg Driver and Nadine Westphal
- Rolf Blädel & Dieter Siemens, Helgoland
- Prof. Dr. Rory P. Wilson, University of Wales, Swansea
- Jens-Uwe Voigt, Driesen & Kern, Bad Bramstedt
- The crew of the vessel *Saibling*, Office for Rural Areas, Husum
- All volunteers who helped during field work

References

Adelung D, Liebsch N, Wilson RP (2004). In MINOS Endbericht (2004): Marine Warmblüter in Nord- und Ostsee: Grundlagen zur Bewertung von Windkraftanlagen im Offshore-Bereich. Nationalpark Schleswig-Holsteinisches Wattenmeer, 2, pp. 335-418

Behrends G (1985). Zur Nahrungswahl von Seehunden (*Phoca vitulina*) im Wattenmeer Schleswig-Holsteins. Z. Jagdwiss. 31:3-14

CWSS (2006): Common Seals in the Wadden Sea in 2006: Puzzling Results. Wadden Sea Newsletters 2006-1. www.waddensea-secretariat.org

Dietz R, Teilmann J, Hendriksen OD, Laidre K (2003). Movements of seals from Rødsand seal sanctuary monitored by satellite telemetry. Relative importance of the Nysted Offshore Wind Farm area to the seals. NERI Technical Report No. 429. Internet version: http://www.dmu.dk/1_viden/2_Publikationer/_fagrapporter/rapporte/FR429.pdf

Hasselmeier I (2006). Evaluation of Blood Tests to assess the Health Status of Harbour Seals (*Phoca vitulina vitulina*) of the German North Sea, PhD-Thesis Univ. Kiel

Jeffries SJ, Brown RF, Harvey JT (1993). Techniques for capturing, handling and marking Harbour seals. Aquat. Mamm. 19:21-25

Kooyman GL (2004). Genesis and evolution of bio-logging devices: 1963-2002. Mem. Natl. Inst. Polar Res. 58, pp. 15-22

Leonhard SB & Petersen J (2006). Benthic communities at Horns Rev before, during and after construction of Horns Rev Offshore Wind Farm. Final report to Vattenfall A/S Bio/Consult, Aarhus, Denmark

Liebsch N (2006). Hankering back to ancestral pasts: constraints on two pinnipeds, *Phoca vitulina* & *Leptonychotes weddellii* foraging from a central place. PhD-Thesis, Univ. Kiel

Liebsch N, Wilson RP, Adelung D (2006). Utilisation of time and space by Harbour seals (*Phoca vitulina vitulina*) determined by new remote sensing methods. In: von Nordheim H, Boedeker D, Krause J. (Eds.): Progress in Marine Conservation in Europe, Springer, pp. 179-188

Mitani Y, Sato K, Shinichiro I, Cameron MF, Naito Y (2003). A method for reconstructing three-dimensional dive profiles of marine animals using geomagnetic intensity data: results from two lacting Weddell seals. Polar Biol. 26: 311-317

Mohr E (1952). Die Robben der europäischen Gewässer. P. Schöps, Frankfurt/Main, pp. 283

Nørgaard N (1996). Haulout-behaviour, movements, foraging strategies and population estimates of Harbour seals (*Phoca vitulina*) in the Danish Wadden Sea. PhD-Thesis, Uni. Aarhus

NPA & UBA (Eds.)(1998). Umweltatlas Wattenmeer, 1. Eugen Ulmer Verlag, Stuttgart, 270pp.

Orthmann T (2000). Telemetrische Untersuchungen zur Verbreitung, zum Tauchverhalten und zur Tauchphysiologie von Seehunden (*Phoca vitulina vitulina*) des Schleswig-Holsteinischen Wattenmeeres. PhD-Thesis, Univ. Kiel

Reijnders PJH (1992). Retrospective population analysis and related future management perspectives for the Harbour seal, *Phoca vitulina*, in the Wadden Sea. In: Dankers N, Smit CJ, Scholl M (Eds), Proceedings of the 7th International Wadden Sea Symposium, Ameland, The Netherlands, 22-26 Oct. 1990,. Neth. Inst. Sea Res., Publ. Ser. 20:193-197

Schreer JF, Kovacs KM, O'Hara Hines RJ (2001). Comparative diving patterns of pinnipeds and seabirds. Ecol. Monogr. 71(1):137-162

Tollit DJ, Black AD, Thompson PM, Mackay A, Corpe HM, Wilson B, Van Parijs SM, Grellier K, Parlene S (1998). Variations in Harbour seal *Phoca vitulina* diet and dive-depth in relation to foraging habitat. J. Zool., London 244, pp. 209-222

Tougaard J, Ebbesen I, Tougaard S, Jensen T, Teilmann J (2003). Satellite tracking of Harbour Seals on Horns Reef. Use of the Horns Reef windfarm area and the North Sea. Report to Techwise A/S March 2003, Syddansk Universitet

Tougaard J, Tougaard S, Jensen RC, Jensen T, Teilmann J, Adelung D, Liebsch N, Müller G (2006). Harbour seals at Horns Reef before, during and after construction of Horns Reef Offshore Wind Farm. Final Report to Vattenfall A/S. Biological papers from the Fisheries and Maritime Museum, No. 5, Esbjerg, Denmark

6 Too loud to talk? Do wind turbine-related sounds affect harbour seal communication?

Klaus Lucke, Janne Sundermeyer, Jörg Driver, Tanja Rosenberger, Ursula Siebert

Zusammenfassung

Die Errichtung von Offshore Windenergieanlagen (WEAs) in der deutschen Nord- und Ostsee stellt einen technischen Eingriff in den Lebensraum der Seehunde und Kegelrobben dar. Einer der wichtigsten Aspekte in diesem Zusammenhang ist die Erzeugung intensiver akustischer Emissionen bei der Rammung der Anlagenfundamente. Ein gesundes Gehör spielt bei Robben zwar nur eine untergeordnete Rolle bei der Nahrungssuche, aber diese Tiere nutzen akustische Signale und entsprechend auch ihr Gehör sowohl unter Wasser als auch an der Luft zur Kommunikation. Speziell die Bindung zwischen Mutter und Jungtier während der Säugephase sowie die territoriale Abgrenzung und Partnerfindung erfolgt bei Robben durch akustische Signale. Die WEA-bedingten Geräusche könnten zu einer Maskierung dieser Signale führen und somit sowohl für einzelne Robben als auch auf Populationsebene schwerwiegende Folgen haben. Eine audiometrische Studie an wilden und trainierten Seehunden soll Grundlagen für eine Beurteilung dieser möglichen Auswirkungen liefern.

Als Untersuchungsmethode wurde die Messung akustisch evozierter Potenziale (AEPs) gewählt. Nach einer Verfeinerung der AEP-Methode konnte an einem trainierten Seehund in der Seehundstation Friedrichskoog ein Audiogramm für Luftschall gemessen werden. Zur erfolgreichen Durchführung der entsprechenden Messungen an wilden Seehunden war es erforderlich, die Tiere zu immobilisieren, um die starken Abwehrreaktionen der Tiere gegen ihre Handhabung während der Untersuchung zu vermeiden. Insgesamt konnten so fünf in der Seehundstation rehabilitierte Heuler und drei freilebende Seehunde audiometrisch erfolgreich getestet werden.

Während die Hörschwellenwerte des trainierten Seehundes teilweise erhöht sind, zeigen die Ergebnisse der wilden Seehunde für Luftschall eine sehr gute Übereinstimmung mit publizierten Hörkurven von Seehunden aus dem Nord-Pazifik und -Atlantik. Die Nutzung der bisher veröffentlichten Daten zur Hörempfindlichkeit von Seehunden unter Wasser zur Beurteilung der Auswirkungen der WEA-Geräusche auf die Kommunikation hiesiger Seehunde scheint entsprechend zulässig. Dabei zeigt sich, dass die Rammimpulse über eine Entfernung von >100km für die Tiere wahrnehmbar sein werden und selbst noch in großer Distanz zur Rammung die Kommunikation der Seehunde maskieren können. Das Maskierungspotential der Betriebsgeräusche wird aufgrund seiner akustischen Eigenschaften als gering eingestuft.

Abstract

The construction of wind turbines in the offshore areas of the German North and Baltic Seas represents a grave intrusion into the habitat of harbour seals and grey seals. One of the most important aspects in this context is the emission of intense acoustic impulses during the ramming of the turbine foundations. While hearing is of secondary importance in seals for locating prey, they use acoustic signals, and hence their hearing, for communication underwater and in air. Acoustic signals are used to keep contact between mother and pups during the nursing phase as well as for territorial display and to attract mating partners. The wind turbine-related sounds could mask these signals and severely impact seals on an individual basis but also on a population level. An auditory study was conducted on trained and wild seals to provide baseline data for the assessment of these effects.

The experiments were conducted by measuring the auditory evoked potentials (AEPs) in the seals. Subsequent to a refinement of the method an audiogram for in-air sound stimuli was measured in a trained seal at the Seal Centre in Friedrichskoog, Germany. To successfully apply this approach to wild seals, the animals had to be immobilised to avoid the strong aversive reaction of the seals to being handled. Five seal pups rehabilitated at the seal station and three free-ranging seals were tested for the hearing sensitivity.

While the hearing thresholds of the trained seals were elevated at in some frequencies, the results for the wild seals are in agreement with previously published hearing data from other studies on harbour seals from the North Pacific and Atlantic. The use of previously published data on the underwater hearing sensitivity in harbour seals was therefore considered acceptable to assess the effect of wind turbine related sounds on their communication. This analysis reveals that the ramming impulses will be detectable by seals over a range of >100km and is likely to mask their communication over wide ranges, too. The potential for masking from the operational noise of wind turbines was considered to be comparably low due to its acoustic characteristics.

Background

Harbour seals and grey seals inhabit the same waters as harbour porpoises, focus of the previous chapters. They are living in the same acoustic environment and faced with the same increase in anthropogenic use of their habitats as whales. Harbour seals and grey seals are the only two seal species breeding in German waters and, like harbour porpoises, have an acute sense of hearing. While seals are not relying on their hearing as much as harbour porpoises, they are confronted with a wide range of potential effects of the planned construction of wind turbines in the German North and Baltic Seas, too. The effects can range from subtle behavioural reactions to activities related to the operation or maintenance of wind turbines to long-lasting habitat loss or even physical injury due to the intense sound emissions during construction of the wind turbine foundations.

Just as for the harbour porpoises, published information on the hearing sensitivity and related auditory parameters is scarce for the two seal species. A single auditory study has been conducted on grey seals so far (Joyce & Ridgway 1975) using an invasive technique to measure cortical auditory evoked potentials. A number of psychophysical hearing tests were conducted on harbour seals over the past decades, both in air and underwater (Møhl 1968, Turnbull & Terhune 1990, Terhune 1991, Kastak & Schusterman 1998). With an increasing number of auditory studies using auditory evoked potentials (AEP) to measure the hearing sensitivity in cetaceans, this technique has recently also been applied to harbour seals (Wolski et al. 2003). The studies revealed that harbour seals have good auditory sensitivity in air over a wide functional hearing range from at least 0.1kHz into the ultrasonic range at

30kHz (Fig. 1). It should be noted that the experiments conducted within the scope of the MINOS+ project were focussed entirely on harbour seals. While knowledge on auditory sensitivity in grey seals remains scarce, the harbour seal data may provide a baseline for assessing the potential auditory effects of anthropogenic noise in general, and wind turbine-related sounds in particular, on this species.

Harbour seals are able to combine a variety of sensory cues to navigate underwater and find prey, most likely along with a spatial memory to orient underwater. They mainly use hydrodynamic and tactile stimuli to forage and navigate (Dehnhardt et al. 1998, 2001; Schusterman et al. 2004), while visual and olfactory sensory inputs are of secondary importance for this function. Since harbour seals are well adapted to an amphibious lifestyle in many aspects of their anatomy and physiology, it is not surprising that these animals also evolved an acute sense of hearing for underwater sounds. Their underwater hearing range is even wider than for airborne sounds. They are able to perceive sounds from below 0.1kHz to at least 128kHz.

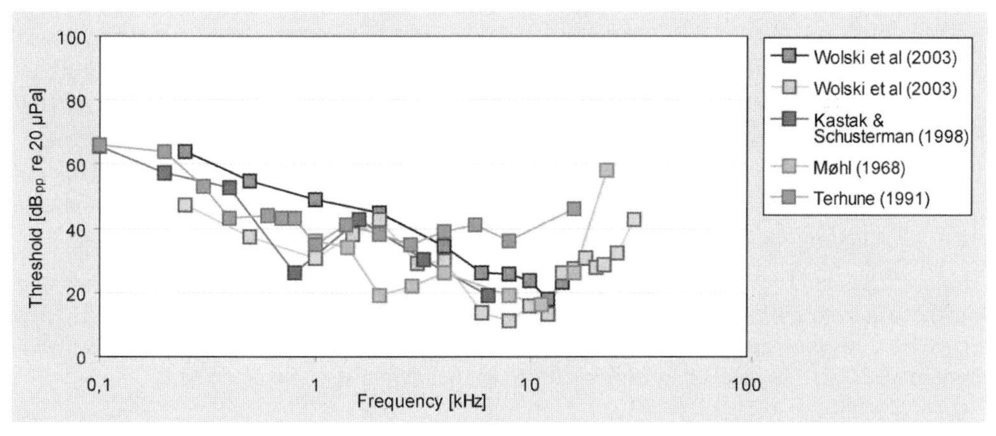

Figure 1: In-air hearing threshold data on harbour seals from several behavioural auditory studies (results from different studies are colour-coded). The coloured squares indicate the hearing threshold measured at a particular frequency, indicating the minimum sound pressure level required by the animals to perceive an acoustic signal at this frequency. Threshold levels are given in 'Decibel peak-to-peak' (dBpp) and referenced against 20µPa, the standard reference pressure in air-borne acoustics.

Acoustic signals are actively (vocalization) and passively (listening) used by seals for communication both in air and underwater. Like all other 32 pinniped species, harbour seals produce sound in air and underwater. There is one significant commonality between all phocid seal species in comparison to other marine mammal species: the apparent structural complexity of their vocalization and degree to which they use sounds actively in communicative contexts. In general, vocalization is used for mother-infant recognition, male-male competition and individual recognition. Calls attracting females to their pups, and vice versa, are marked by sharp onsets and are frequency- and amplitude-modulated. Threats and alarm calls are structurally similar across species and sexes and are found in all species. Males have individually recognizable calls produced in the context of aggressive exchanges. Those are often short, repetitive broadband pulses with rapid onset. The main function of pinniped vocalization is to elicit attention of recipients and guide their behaviour in a way that is beneficial to the signaller (Schusterman et al. 2001). Land-breeding seals, such as the grey seal, congregate seasonally on crowded rookeries and use vocalization as the major tool to communicate with each other. Aquatic breeding seals, like the harbour seal, are rarely producing sounds in air (exceptions: pups, occasional guttural threats). In contrast to their behaviour on land, they develop unusual and complex underwater vocal repertoires. The variability in pinniped vocal signaling can be assessed at the level of the individual (age, gender), emotional state, geographic location and species (Schusterman, Van Parijs 2003).

Harbour seals have long been thought to be a silent species without much vocalization both in air and underwater. However, recent studies have shown that males are vocalizing underwater during the mating season (Van Parijs et al. 1999). Underwater vocalization in males may be used to attract females but also in the context of territorial and competitive behaviour with other males (Bjørgesæter et al. 2004, Hanggi & Schusterman 1994, Van Parijs et al. 1999, 2000a, 2000b).

Female harbour seals are widely distributed during mating season at sea, and male harbour seals do not monopolize females during the mating season, as has been observed in other phocid seals (Thompson et al. 1994). It is also known that vocalization in male harbour seals varies in space and time in relation to diurnal or tidal patterns (Van Parijs et al. 1999) and there is evidence of variations in vocalization between different colonies (Bjørgesæter et al. 2004).

The known frequency range of the underwater roar of harbour seals is 0.4 to 4.3kHz, with dominant frequencies between 0.4 to 0.8kHz (Hanggi & Schusterman 1994, Richardson et al. 1995). Typical source levels of underwater are 130dB re 1μPa (Dudzinski et al. 2002).

Underwater sounds produced during mating season are often preceded by fore-flipper slaps on the water surface, which cause short click-like sounds (Richardson et al. 1995). These slaps are 2 ms long and are estimated to reach source levels of 186-199dB_{pp} re 1 μPa (Wahlberg et al. 2002). The seals may also perform this behaviour to show aggression.

Problem

Seals may not be vitally dependent on their auditory system to survive as adult individuals, but a functional hearing system and the unrestricted perception of biologically meaningful sounds is a key issue for seal pups during the nursing phase whenever they are separated from their mothers. Reproductive success may be affected by sound-induced disturbance and thus would have implications on a population level.

Beside the aforementioned behavioural disruptions, acoustic signals may lead to short- or long-term displacement from valuable feeding or resting habitats and interrupt or prevent successful congregation of mating pairs. In extreme situations, intense sound may lead to injury in seals. At all levels of perceived intensity, noise will cause stress in the animals, with unknown short- and long-term effects.

Important aspects in need of assessment in conjunction with the construction of wind farms in or near seal habitats are the potential effects of wind turbine-related sounds on seal communication and the tolerance limit of their auditory system. While these two aspects can in principle be studied in captive as well as in wild seals, it is essential to monitor the health status of the hearing system in wild populations. Deterioration in their auditory health may be detected and causes identified. Meaningful mitigation strategies could be developed on the basis of this information.

Consequently, one aim of the acoustic studies on seals within the project MINOS+ was to develop and establish a technique which would allow us to conduct auditory measurements on trained seals in captivity as well as on wild seals. This method should then be used to develop a data basis on the hearing sensitivity of wild seals and compare these data with hearing thresholds acquired in captive animals. Given the comparability of data from both groups, experiments with the captive animals were envisaged to test the effects of wind turbine-related sounds on their communication and the tolerance limit of their hearing system to sound exposures. Just as for harbour porpoises, this information on seal hearing is required to assess the potential effect of construction and operation of the planned wind turbines and define noise exposure criteria which may be species-specific and thus different from present criteria.

Study design

Measuring auditory evoked potentials (AEPs) was considered to be the optimum method for the planned experiments. However, the study design for the seal experiments would in many respects have to be different from the harbour porpoise studies. First, the technique should be used for measurements in air and underwater. The different physical properties of the two media would need to be taken into account for stimulus presentation and AEP measurements. Second, the experiments should be conducted on captive and on wild seals, which would most likely involve strong aversive reactions of the subjects once they were caught and restrained. Another parameter which had to be considered in the study design is the different anatomy of the hearing system in seals compared to harbour porpoises. While the outer ear canal is no longer functional in whales, seals perceive sound in air mainly by airborne transduction via the outer ear canal. While it closes as soon as they dive, they are also capable of closing this canal while they are on land, thus impeding the perception of sound.

The studies on captive harbour seals were conducted at the Seal Centre Friedrichskoog, Germany. Two of their animals were trained to participate in the studies and *Deern* at 22 years of age the oldest female of the group, proved to be the most reliable animal to take part in the studies.

The experimental setup was based on the same principle of measuring the auditory evoked potentials (AEPs) as described in chapter 4. A workstation (Tucker-Davis Technologies – TDT, System 3) was used to generate the sound stimuli, which were presented either via speakers (Magnat Quantum 501 or Monacor LSP 60) or via headphones (Beyerdynamic DT 48 A). The evoked potentials were measured with surface or needle electrodes (NIHON-Kohden) and fed into an input module (TDT RA 4 LI) and a biopotential amplifier (TDT RA 4 PA), and subsequently into the TDT workstation for real-time display, further analysis and data storage. The recorded potentials were bandpass-filtered between 300Hz and 3kHz. Under normal measurement conditions 500 AEPs were averaged to achieve one data point.

As soon as *Deern* was trained well enough to start the experiments, a comparison of different electrodes and a mapping for the best placement for these electrodes were conducted (Sundermeyer 2006). While the neuronal response was stronger if measured with needle electrodes, the animal showed strong aversive reactions as soon as these electrodes were applied. Surface electrodes, in contrast, proved to be less aversive and still provided signal quality sufficient for the purpose of this study.

After the optimum placement of the electrodes was determined, different types of sound stimuli for the AEP experiments were tested (Sundermeyer 2006). Short tone pips consisting of cosine-gated sine waves yielded the best results. As the frequency specificity of the signals and the amplitude of the evoked neuronal responses are negatively correlated, the best combination of the two factors had to be determined. In practice, pips consisting of 8 cycles (i.e. a bandwidth of 0.2 octaves, Reichmuth et al. 2007) were most often used for frequencies above 1kHz, while AEPs could only be evoked at sufficient levels for lower frequencies if the number of cycles was reduced to 3 (bandwidth ~0.8 octaves). The loss of frequency specificity was regarded as an acceptable trade-off for successful threshold measurement at the low frequencies as these were of special interest in the context of the wind turbine-related sounds.

The acoustic masking (i.e. the impaired perception of a sound due to the presence of another sound) of the AEP stimuli can best be avoided by reducing the surrounding sound (as in a sound proof chamber) or by dampening the environmental noise through the use of headphones. The initial experiments involving headphones for stimulus presentation showed inconsistencies in the recorded AEPs, possibly caused by the seal's closing her outer ear canals – either as a reflex or voluntarily – thus reducing the perceived sound pressure level by approximately 40dB (estimate based on the resulting differences in threshold levels from consecutive measurements at the same frequency). Therefore, speakers were used for the in-air hearing tests with *Deern*, with the consequence that background noise could no longer be effectively reduced. Instead, the experiments were preferably conducted during quiet periods (Fig. 2).

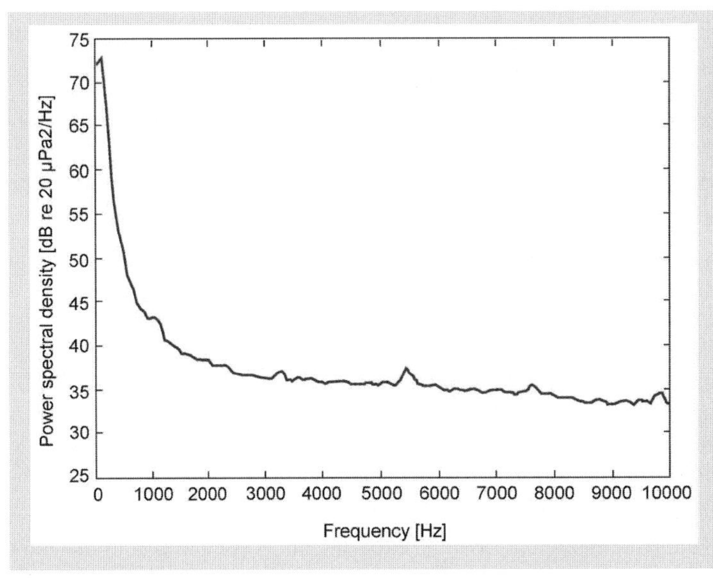

Figure 2: Background noise level (plotted as power spectral density) recorded in air at the seal's position during the AEP measurements at the Seal Centre Friedrichskoog in quiet conditions.

Physiological masking of AEPs by myogenic artefacts was another limiting factor considered to cause increased thresholds. Consequently, *Deern*'s training was focused on preventing her from showing any unnecessary muscular activity during the experiments. This was partially achieved by separating her spatially from the other seals during the experiments, thus avoiding any physical disturbance. She was trained to rest her head on a chin-cup as this had proven to reduce her head movements and the muscular tension in her neck from trying to hold her head up slightly.

Access to wild seals was usually possible during seal catches in the German Wadden Sea area, which were routinely conducted twice per year as part of a health monitoring programme. These catches take place on sandbanks, and seals can be physically restrained and tested under veterinary control for a maximum of up to 1.5 hours. The attempts to measure AEPs in restrained seals however yielded no reproducible results.

The need to find alternative ways to test wild animals brought the focus back to the work at the Seal Centre in Friedrichskoog. The Seal Centre, in addition to providing environmental education and research on seals, is also one of two rehabilitation facilities for seal pups at the German North Sea coast, thus offering unique access to wild seals for integration into the studies.

After different techniques to calm down or restrain the animals had been tested without success, it was decided to immobilise the animals for the duration of the measurements, which were conducted indoors at the Seal Centre. Several different anaesthetics have been successfully tested and used during clinical treatments in seals by specialists around the world. Based on their experience a combination of drugs was applied to one of the seals. The immobilisation prevents almost all voluntary muscular activity in the animal, and the AEP measurements instantly showed clear and reproducible results. The successful application of this method was first confirmed and then used routinely for all AEP measurements in wild seals. All seals were continuously observed by an experienced veterinarian during the experiments as well as for at least one hour afterwards until they were found to again behave normally.

Results

The measurements to find the best placement for the electrodes revealed that the active electrode had to be placed on a dorsal line along the body at a position 2cm in front of an imaginary line ('ear line') between the two ear openings. As the system was not grounded via water, as in the AEP studies on the harbour porpoise, a ground electrode was applied. Its best position was found to be 15cm behind the ear line, while the best results where achieved with the reference electrode at a position 25cm behind the ear line.

A full audiogram was measured in *Deern* in air with all stimuli presented via speakers (Fig. 3). The results are plotted in dB_{pp} to allow for comparison with previously published data.

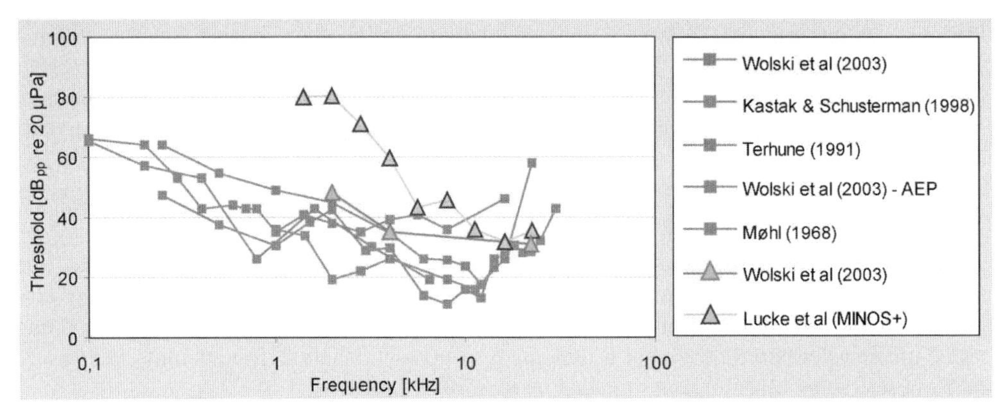

Figure 3: In-air hearing threshold data of a female harbour seal acquired by measuring neuronal responses (AEPs) to the repeated presentation of short cosine-gated sine waves. Blue triangles represent values measured in this study, green triangles show results from another AEP-hearing study on harbour seals and grey squares show data from behavioural auditory studies (see Figure 1) for comparison.

Deern's auditory sensitivity was measured at frequencies ranging from 1.4kHz to 22.4kHz at half octave intervals. Below 5.6kHz, the resulting hearing thresholds were elevated by up to 40dB compared to similar data from behavioural studies and the only other AEP study on harbour seals. At 5.6kHz and above, however, her hearing sensitivity was almost identical to results of other studies. The typical mammalian U-shaped audiogram is only reflected in the slight increase in *Deern*'s hearing threshold at 22.4kHz. The lowest threshold, i.e. her best sensitivity, was measured at 16kHz. Apart from this highest frequency, her hearing sensitivity improved with increasing frequency. This is in contrast to the results published for other harbour seals, which had their best in-air hearing sensitivity between 8kHz and 11.2kHz (Møhl 1968, Terhune 1991, Kastak & Schusterman 1998).
If analysed in combination with the prevailing background noise (Fig. 4), hearing thresholds are still clearly elevated above the background noise level at frequencies below 5.6kHz. At frequencies above 5.6kHz, they are in good agreement with the measured noise level.

Hearing thresholds were also obtained for five of the harbour seal pups rehabilitated at the Seal Centre. The animals were less than 4.5 months old and weighed less than 33kg at the time of the experiments. In contrast to the hearing tests with *Deern*, these measurements were conducted in the seal station's medical laboratory. During the measurements background noise was kept to a minimum to reduce acoustic masking. Resulting hearing thresholds from the five individuals are plotted in Figure 5.
The five harbour seal pups showed identical or comparable levels of hearing sensitivity at some frequencies but also deviations between some individuals of up to 35dB. While the highest levels were measured on average in the female seal *Pv 224 BRI* another female, *Pv 275 ANN*, had the lowest hearing thresholds on average. Except for the lowest frequencies (0.7kHz to 1.4kHz), the audiograms of all five harbour seal pups showed a comparable trend and revealed a typical mammalian U-shape. A slight to medium increase at 11.2kHz was measured in three of the five animals. All five audiograms are in accordance with the hearing threshold data from adult harbour seals published so far (see inserted graph in Figure 5).

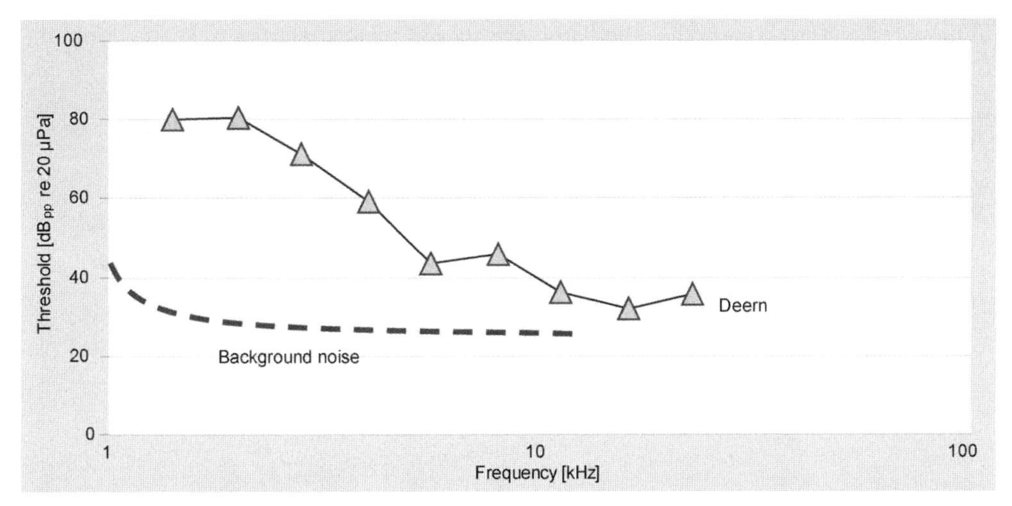

Figure 4: In-air hearing threshold data of a female harbour seal (cf. Figure 3) plotted in combination with the spectral density (schematic) of the background noise measured up to 10 kHz in air at the animal's position in quiet conditions (cf. Figure 2).

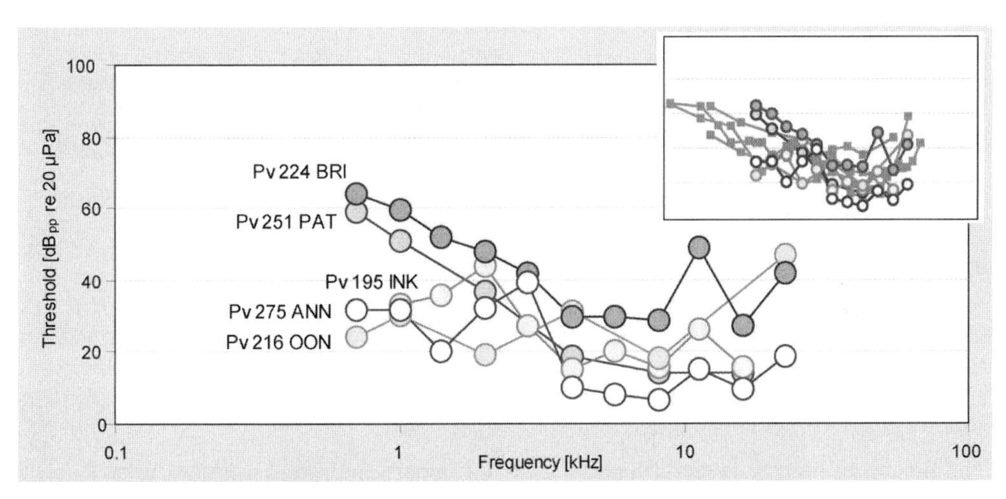

Figure 5: In-air hearing threshold data of five harbour seal pups (results from different individuals are colour-coded) acquired by measuring neuronal responses (AEPs) to the repeated presentation of short cosine-gated sine waves. The measurements were conducted indoors in quiet conditions and all animals were immobilised during the measurements. The small inserted figure shows the resulting thresholds for the harbour seal pups in comparison with hearing thresholds (grey lines) acquired in behavioural auditory studies on adult harbour seals (cf. Figure 1).

The data gathered from the wild animals cover the frequency range between 1kHz and 16kHz, but the amount of data achieved so far is still insufficient to plot a complete audiogram of a free-ranging harbour seal. While the data from the three animals differ slightly, their variation is smaller than in the harbour seal pups tested at the Seal Centre. Comparison of the threshold levels from the wild harbour seals and the seal pups reveals no general difference (see inserted graph in Figure 6).

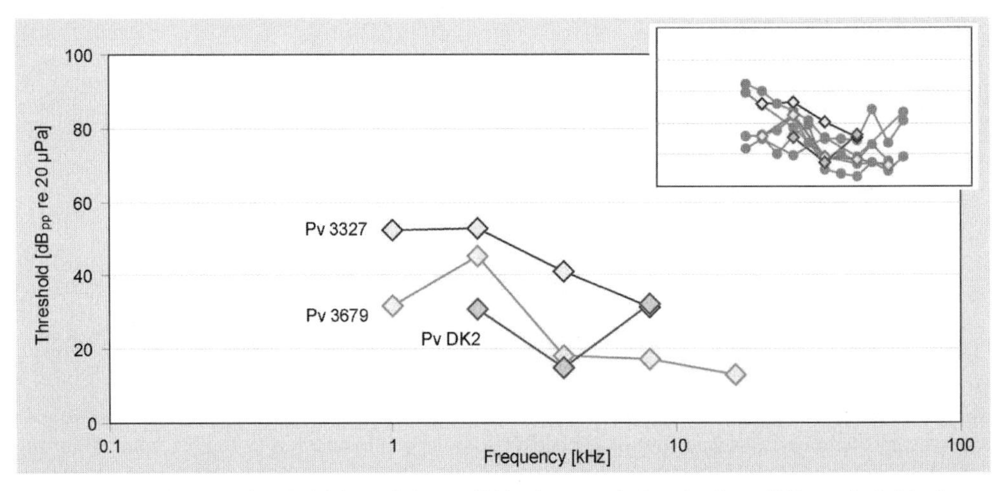

Figure 6: In-air hearing threshold data of three wild harbour seals (results from different individuals are colour-coded) acquired by measuring neuronal responses (AEPs) to the repeated presentation of short cosine-gated sine waves. The measurements were conducted on a sandbank in the German Wadden Sea under comparatively quiet conditions, and all animals were immobilised during the measurements. The small inserted figure shows the resulting thresholds for the wild seals in comparison with hearing thresholds (grey lines) acquired in behavioural auditory studies on adult harbour seals (cf. Figure 5).

A comparison of emitted sounds with underwater hearing sensitivity of seals is necessary to assess the potential impact of wind turbine-related sound emissions on the communication of harbour seals.

As long as AEP-data on the underwater hearing thresholds are not available, comparable data from behavioural studies have to be used as a baseline. Those are plotted together with the sound pressure level of ramming impulses recorded at two different distances from the piling site during the installation of a small pile for a research platform in the North Sea (Amrumbank West, data provided by Betke et al. 2006 – pers. comm.).

The functional hearing range of harbour seals for underwater sound stretches from the very low frequencies (below 100Hz) to 128kHz and probably above. Their best sensitivity varies individually or depending on the design of the behavioural studies between 8kHz and 32kHz.

The sound pressure level of the piling impulses as recorded near the piling site reaches its highest levels at frequencies of approximately 500Hz. While the hearing sensitivity of seals is not the best at these low frequencies, the sound pressure level of the ramming impulses is >100dB above their threshold. Measured at a distance of 33km from the sound source, the main energy in the frequency spectrum of the piling impulses is shifted towards the higher frequencies, i.e. lower frequencies were more attenuated over distance than higher frequencies. Even at this distance, the maximum sound pressure level is still 40 to 50dB above the threshold of the seals at the given frequencies.

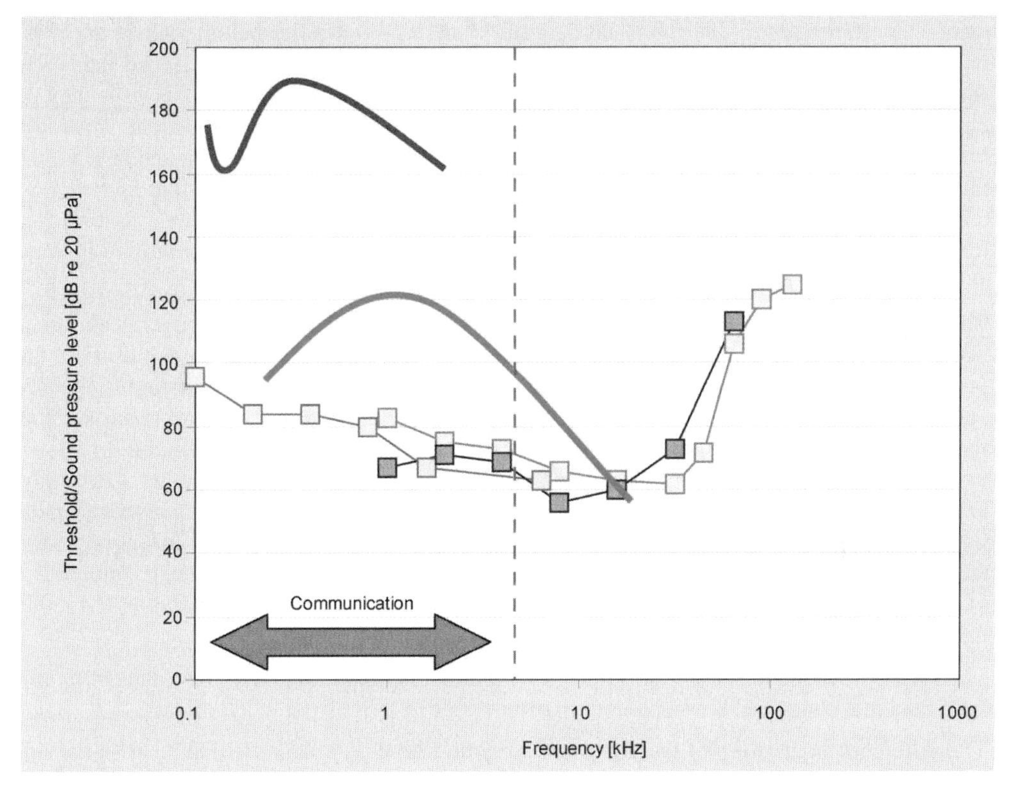

Figure 7: Hearing thresholds of harbour seals for underwater sound measured in different behavioural studies (results from different studies are colour-coded; blue squares: Kastak & Schusterman 1998; orange squares: Terhune 1988; yellow squares: Møhl 1968). The coloured squares indicate the hearing thresholds measured at a particular frequency, indicating the minimum sound pressure level required by the animals to perceive an underwater acoustic signal at this frequency. The red line indicates the spectral distribution of the sound pressure level of a ramming impulse (1/3 octave band analysis; schematic) as measured in the vicinity of the piling site (based on data measured at 36m distance to the piling of a 1.5m diameter pile). The green line indicates the equivalent data (1/3 octave band analysis; schematic) for an impulse measured at 33km distance. Threshold levels and sound pressure levels are referenced against 1µPa, the standard reference pressure in underwater acoustics. The dashed grey line indicates the upper limit of the frequency range (<5kHz) of most communication signals (orange arrow) of harbour seals.

Conclusion / Discussion

The results of this study show that the AEP technique is suitable for auditory measurements on both free-ranging and captive harbour seals. This method is now well established and should be used in the future to increase the data basis on auditory sensitivity in this species.

Adequate stimulus types and sound sources as well as electrode placement must be carefully selected. The elevated hearing thresholds resulting from the study on the trained seal indicate that acoustic masking by background noise is a limiting factor when these factors are considered. The differences to the published data of >30dB at frequencies below 5.6kHz are most likely linked to the comparatively high level of background noise which exceeded the levels shown in Figure 2 during the auditory measurements on *Deern* on average. This issue may be eliminated in future studies by conducting the experiments in a soundproof room. Thus, the resulting thresholds for *Deern* should be considered masked hearing thresholds, at least for the low frequency range. Nevertheless, using trained seals is the only feasible approach to testing the underwater hearing sensitivity and related aspects in these animals. Limiting factors, such as background noise and the animal's muscular activity, must be minimised in these studies as much as possible and carefully monitored if possible.

The use of anaesthetics to reduce myogenic artefacts and eliminate conscious muscular control over the outer ear canal was successfully applied in this study. Under immobilisation, the outer ear canal remains open and headphones can be used to present the AEP-stimuli, thereby attenuating environmental noise by e.g. 12dB. If conducted in a quiet environment such as the seal station's medical room, acoustic masking cannot be entirely ruled out, but as shown by the comparison with the published data from behavioural studies and another AEP study, the resulting thresholds can be regarded as absolute thresholds. These measurement conditions therefore seem to be almost ideal for an auditory study in seals. Immobilisation of the animals is the key for successful in-air measurements of AEPs in untrained seals, but control of all other parameters discussed is equally important to achieving reliable and reproducible data.

The results show that, while a fairly small number of animals was tested, the data from animals held in a controlled environment are comparable to those from wild animals.

The difference in duration of the presented acoustic stimuli is a factor which must be considered in the comparison of threshold data from behavioural studies and AEP studies. The duration of stimuli used in AEP studies is usually in the range of a less than ten milliseconds while stimuli in behavioural studies are mostly several hundred milliseconds or even seconds long. In normal mammalian hearing, acoustic perception improves - i.e. the threshold level decreases - if the duration of a signal increases up to the duration of the integration time of its hearing system (involving higher cortical levels which process the information). First results (Mulsow et al. 2007) indicate that the integration time in seals is in the range of 1.6ms to 5.3ms, therefore overlapping with the range of stimulus durations used in this study. If stimulus durations fall below the integration time of the auditory system of seals, resulting thresholds would need to be corrected (i.e. reduced) by a factor of 3dB per doubling of time.

Wind turbine-related signals are mainly emitted during the construction phase and are considerably more intense underwater than in air. To assess the potential impact of these underwater sounds on seal hearing or communication, corresponding underwater hearing sensitivity must be considered. Based on the good agreement between hearing thresholds from behavioural and AEP-studies for in-air signals, application of the published underwater hearing data seems valid for this purpose even though they have been measured on animals from the North Pacific or Northeast Atlantic. These animals thus inhabited an acoustically different environment, and the health status of their acoustic system might be different compared to animals from the North Sea. Due to the overlap in the achieved in-air hearing thresholds this difference was considered negligible and to be outweighed by the lack of AEP data on the underwater hearing sensitivity in seals from the North Sea.

A comparison of the spectra and sound pressure levels of sounds emitted during the construction of wind turbines with the underwater hearing sensitivity of harbour seals (Fig. 7) indicates that seals will most likely be able to perceive these signals at a distance of a hundred kilometres from the piling site or more. This estimate may vary by some tens of kilometres (in both directions) because propagation loss can vary in relation to water depth, bottom topography and related physical parameters.

In the context of seal communication, it is important to note that the main energy of the piling impulses is concentrated in the low frequency range, thus overlapping with the frequency range of harbour seal communication signals. As long as the perceived levels of the piling noise are above the perceived level of a communication signal or any biologically meaningful signal, the seals will be unable to detect those signals. The perception of sound in the presence of masking sound does however not depend on the absolute hearing threshold alone. It is also determined by other auditory parameters like the critical ratio. The critical ratio defines the level (in dB) at which a sound (a pure tone - per definition) can be detected above the background or, as in this case, the masking noise level.

Terhune (1991) found that for white noise the critical ratio in harbour seals was 24dB at 0.25kHz, 15dB at 0.5kHz and 21dB at 1kHz. Given that these values apply for harbour seals irrespective of their geographical origin, the seals in the North Sea would not be able to detect any communication sounds as long as the perceived levels of these signals do not exceed the noise from the piling impulses by this amount. Given a propagation loss of 16log r (with r = the distance in m, a typical value for the North Sea), the masking range would be at least double if this parameter applies. Because piling impulses are neither pure tones nor true broadband signals, it remains unclear to which extent this effect exists for piling impulses as long as it has not been specifically tested.

Temporal characteristics of masking noise, in addition to spectral information, are another factor requiring consideration. Near the source, piling impulses are less than 100ms long and usually are repeated at rates between 2 and 30 impacts per minute, i.e. at intervals between 2 and 30s. A 100ms long impulse would hardly be able to mask even the shortest communication signals of seals as e.g. probably with the exemption of the fore-flipper slaps on the water surface. But with increasing distance from the sound source, the piling impulses are lengthened due to reflections at the boundaries. The duration of impulses measured during piling of the foundation for a research platform in the North Sea, for example, increased from 100ms near the source to 400ms at 17km distance (Betke, pers. comm.). As the received levels at such distances are still high enough to mask communication signals, such long piling impulses clearly have the potential of masking any sounds in a temporal respect, especially if they are repeated at a high rate.

Such masking of inter-animal communication or biologically meaningful sounds would have negative consequences for individual seals, if the contact between a mother and her pup is disturbed or impeded, as well as for the population, e.g. if successful reproduction is impaired.

The potential masking effect of the operational noise of wind turbines will be considerably lower though this noise will be emitted continuously. Most of this low-frequency noise is produced at the turbine itself (especially in the gear box) and transmits into the water mainly via the immersed section of a wind turbine's foundation. The source levels of this type of noise are much lower compared to the construction noise (130 to 140dB re 1μPa at maximum, Madsen et al. 2006), and the most prominent peaks in the spectral density distribution are narrow-band tonal components so that the risk of acoustic masking is reduced compared to the broader spectral distribution in piling impulses.

Behavioural reactions elicited by the perception of noise are likely to occur. The lack of knowledge, however, does neither allow qualitative nor quantitative assessment. The potential physical effects of intense sound on the hearing system of harbour seals has been tested within the scope of the MINOS+ project, but the levels at which a ramming impulse could be reproduced in air were not high enough to cause any threshold shift. These experiments will continue until a noise exposure criterion can be defined for harbour seals.

The strongest need for research currently exists with regard to grey seals as hardly any information exists on their auditory sensitivity. This data gap should be closed in due time to allow for a comprehensive assessment of the potential effects of wind turbine-related acoustic emissions as well as many other anthropogenic sound sources on all marine mammal species in German waters.

Acknowledgments

We would like to thank Svend Tougaard and Tyge Jensen from the Fisheries and Maritime Museum, Esbjerg, Denmark, and the staff of the Marine Rescue Station in Havneby on the island of Rømø, Denmark, as well as all volunteers at the seal catches. Support by Rainer Lüdtke from the Office for Rural Areas, Husum, and the crew of the vessel *Saibling*, was also much appreciated. Parts of the equipment were provided by MAGNAT and the GKSS Research Centre, Geesthacht. All experiments at the Seal Centre and on the sandbanks off Schleswig-Holstein were conducted with permission of the Schleswig-Holstein Ministry of Agriculture, Environment and Rural Areas.

References

Bjørgesæter A, Ugland KI, Bjørge A (2004). Geographic variation and acoustic structure of the underwater vocalization of harbour seals (*Phoca vitulina*) in Norway, Sweden and Scotland. J. Acoust. Soc. Am., Vol. 116 (4):2459-2468.

Dehnhardt G, Hanke W, Bleckmann H, Mauck B (2001). Hydrodynamic trail-following in harbour seals (*Phoca vitulina*). Science, Vol. 293:102-104.

Dehnhardt G, Mauck B, Bleckmann H (1998). Seal whiskers detect water movements. Nature, Vol.394:235-236.

Dudzinski KM, Thomas JA, Douaze E (2002). Communication. P. 248-268. In: Perrin WF, Würsig B, Thewissen JGM (Eds.): Encyclopedia of marine mammals. Academic Press.

Hanggi EB & Schusterman RJ (1994). Underwater acoustic displays and variation in male harbor seals, *Phoca vitulina*. Anim Behav, Vol. 48 (6):1275-1283.

Kastak D & Schusterman RJ (1998). Low-frequency amphibious hearing in pinnipeds: Methods, measurements, noise, and ecology. J. Acoust. Soc. Am., Vol. 103:2216-2228

Madsen PT, Wahlberg M, Tougaard J, Lucke K, Tyack P (2006). Wind turbine underwater noise and marine mammals: Implications of current knowledge and data needs - Review. Marine Ecology Progress Series, Vol. 309:279-295.

Møhl B (1968). Auditory sensitivity of the common seal in air and water. J. Aud. Res., Vol. 8:27-38.

Mulsow J & Reichmuth C (2007). Electrophysiological assessment of temporal resolution in pinnipeds. Aquat. Mamm., Vol. 33 (1):122-131.

Reichmuth C, Mulsow J, Finneran JJ, Houser DS, Supin AYa (2007). Measurement and response characteristics of auditory brainstem responses in pinnipeds. Aquat. Mamm., Vol. 33 (1):132-150.

Richardson WJ, Greene CRJr, Malme CI, Thomson DH (1995). Marine Mammals and Noise. San Diego: Academic Press. 576 pp.

Schusterman RJ, Kastak D, Levenson DH, Reichmuth-Kastak C, Southall BL (2004). Pinniped sensory systems and the echolocation issue. In: Thomas JA, Moss C, Vater M (Eds): Echolocation of Bats and Dolphins. University of Chicago Press, pp. 531-535.

Schusterman RJ, Southall BL, Kastak D, Reichmuth-Kastak C (2001). Pinniped vocal communication. Proceedings of the 17th International Congress on Acoustic Proceedings, Rome, Italy, 2001.

Schusterman RJ & Van Parijs SM (2003). Pinniped vocal communication: an introduction. Aquat. Mamm., Vol. 29.2:177-180.

Sundermeyer J (2006). Untersuchungen zur Optimierung der AEP-Methode an Robben. Diplomarbeit. Institute for Polar Ecology, Christian-Albrechts-University Kiel. In German.

Terhune JM (1988). Detection thresholds of a harbour seal to repeated underwater high-frequency, short-duration sinusoidal pulses. Can.J. Zool., Vol. 66:1578-1582.

Terhune JM (1991). Masked and unmasked pure tone detection thresholds of a harbour seal listening in air. Can. J. Zool., Vol. 69:2059-2066.

Thompson PM, Tollit DJ, Wood D, Corpe H, Hammond PS, Mackay A (1997). Estimating harbour seal abundance and status in an estuarine habitat in northeast Scotland. J. Appl. Ecol., Vol. 34:43-52.

Van Parijs SM, Hastie GD, Thompson PM (1999). Geographical variation in temporal and spatial vocalization of male harbor seals in the mating season. Anim. Behav., Vol. 58:1231-1239.

Van Parijs SM, Hastie GD, Thompson PM (2000a). Individual and geographical variation in display behaviour of male harbour seals in Scotland. Anim. Behav., Vol. 59:559-568.

Van Parijs SM, Hastie GD, Thompson PM (2000b). Display area size, tenure and site fidelity in the aquatic mating male harbor seal. Can. J. Zool., Vol. 78:2209-2217.

Wahlberg M, Lunneryd S-G, Westerberg H (2002). The source level of harbour seal flipper slaps. Aquat. Mamm., Vol. 28 (1):90-92.

Wolski LF, Anderson RC, Bowles AE, Yochem PK (2003). Measuring hearing in the harbour seal (*Phoca vitulina*): Comparison of behavioural and auditory brainstem response techniques. J. Acoust. Soc. Am., Vol. 113 (1):629-637.

Excursus 5: Gulls and Auks

Katrin Wollny-Goerke, Kai Eskildsen, Nele Markones, Stefan Garthe

The **Herring Gull** (*Larus argentatus*) is one of the most well known larger gulls and one of the most common gulls at the coast. It is a big gull with a light grey back and grey wings. The yellow bill with its red spot near the tip of the lower mandible is conspicuous. The legs are flesh-coloured to pink which, in addition to its size, distinguishes the Herring Gull from the Common Gull.

Herring Gulls are widespread along German coasts and breed both at the North Sea and Baltic coasts in dunes, grassy isles, and on shingle beaches in colonies. This coastal species forages predominantly in the nearshore sea areas and the mud flats of the Wadden Sea. The variable diet consists mostly of marine invertebrates but also includes fish, discards from fishing vessels, terrestrial organisms and refuse from dump sites. At sea, Herring Gulls are widely distributed along the coasts. Like in the other gull species, concentrations of Herring Gulls are rather local and short-lived but abundances can be high (see chapter 7). Herring Gulls belong to the migratory species regularly occurring in the German North and Baltic Seas, whose breeding, moulting, resting, migration and wintering areas have to be protected according to the EU Birds Directive.

The grey-white **Common Gull** (*Larus canus*) resembles the Herring Gull but is considerably smaller. Bill and legs are yellow-green and, unlike the Herring Gull, its bill does not have a red spot near the tip.

Common Gulls are less restricted to coastal habitats than Herring Gulls. Foraging areas include, next to the intertidal zone, also terrestrial habitats such as farmland and grassland. Common Gulls have a particularly broad food spectrum, comprised above all of marine and terrestrial invertebrates but also fish, terrestrial vertebrates, fisheries discards and refuse.

In the German North Sea, this gull species is concentrated near the coast in all seasons (see chapter 7), but the distribution area extends also further offshore during winter. In the German Baltic, Common Gulls are particularly widespread during winter. Common Gulls belong to the migratory species regularly occurring in the German North and Baltic Seas, whose breeding, moulting, resting, migration and wintering areas have to be protected according to the EU Birds Directive.

At first sight **Black-legged Kittiwakes** (*Rissa tridactyla*) resemble Common Gulls. However, Kittiwakes have no white spots on the black wingtips and their legs are black.
Kittiwakes are high-sea birds, visiting land only at breeding time. In Germany they breed exclusively on the island of Helgoland. Kittiwakes build their nests in small rock shelters, on cliff ledges and in small grassy patches on cliff sides.
Kittiwakes are surface-feeding seabirds, taking up their prey of the sea surface and the upper decimetres of the water column in flight. During plunge-diving they can reach maximum depths of 1-2 metres. They feed on small pelagic fish and marine invertebrates, utilising offshore areas for feeding (chapter 7).

While during summer and autumn clear distribution centres are evident in the vicinity of Helgoland, the species is evenly distributed throughout the entire EEZ during winter. Distribution patterns are characterised by short-term local concentrations (see chapter 7).
Kittiwakes rarely occur in the German Baltic area, and if so, it is primarily in connection with strong westerly winds.
Kittiwakes belong to the migratory species regularly occurring in the German North Sea, whose breeding, moulting, resting, migration and wintering areas have to be protected according to the EU Birds Directive. Special conservation interest has to be directed to the single breeding site in German waters on the island of Helgoland.

The **Lesser Black-backed Gull** (*Larus fuscus*) is a medium-sized gull with dark grey to black upperparts. It can be distinguished from the Great Black-backed Gull by its smaller size and its yellow legs.

In Germany, large colonies are found along the North Sea coast while only a few pairs breed on the Baltic coast. Lesser Black-backed Gulls regularly occur in German waters from March to early October and are only occasionally sighted during winter. They winter from western Europe down to West Africa.

In the North Sea, the Lesser Black-backed Gull hardly utilises the mud flats of the Wadden Sea area for foraging purposes but feeds almost exclusively at sea at distances of up to 80km from the breeding colonies (see chapter 7). Their diet includes fish and marine invertebrates, with swimming crabs (*Portunidae*) playing an important role at certain times. Fisheries discards are also utilised as a food source. Lesser Black-backed Gulls belong to the migratory species regularly occurring in the German North Seas, whose breeding, moulting, resting, migration and wintering areas have to be protected according to the EU Birds Directive.

Black-headed Gulls (*Larus ridibundus*) are roughly dove-sized but very slim. Their bill is slender and noticeably red. In summer the gull has a conspicuous chocolate-brown hood. In winter this is reduced to a brownish-black spot behind the eye.

The Black-headed Gull is the most common gull species in Europe. It is widespread along German coasts and can also be found in large flocks inland, for example on farmland, near lakes and on rubbish dumps. Black-headed Gulls are distinctly colonial in their breeding. Other seabird species such as terns (*Sternidae*) often breed in the protection of their colonies. The diet often includes plant material and comprises a broad spectrum of marine and terrestrial organisms, above all invertebrates such as earthworms, polychaetes, insects and molluscs.

In the German North and Baltic Seas, Black-headed Gulls occur predominantly in areas near the coast (see chapter 7). They belong to the migratory species regularly occurring in the German North and Baltic Seas, whose breeding, moulting, resting, migration and wintering areas have to be protected according to the EU Birds Directive.

Common Guillemots (*Uria aalge*) belong to the auks (*Alcidae*). They have a characteristic black and white plumage and are truly seabirds, coming only to the coast during the breeding season. They nest in large breeding colonies on steep cliffs. They mostly lay only one egg which is placed on bare rock and is incubated on the bird's feet. At between 18 and 24 days of age the chicks, which are still unable to fly at that time, leave the nest by jumping off the rock ledge at dusk and into the sea. The parents continue to care for and feed the young at sea for several weeks thereafter.

Guillemots feed predominantly on pelagic schooling fish species. They are pursuit divers who skilfully use their short and narrow wings to propel themselves underwater.

In Germany, Guillemots only breed on Helgoland and use the island's vicinity as a feeding ground (see chapter 7). During the winter they occur in many parts of the German EEZ, sometimes at high densities.

In the German Baltic Sea, this auk species occurs primarily during winter and predominantly in the Pomeranian Bight. Only small numbers are found in other areas and seasons. Guillemots belong to the migratory species regularly occurring in the German North and Baltic Seas, whose breeding, moulting, resting, migration and wintering areas have to be protected according to the EU Birds Directive. Special conservation interest has to be directed to the single breeding site in German waters on the island of Helgoland.

7 Small-scale temporal variability of seabird distribution patterns in the south-eastern North Sea

Nele Markones, Stefan Garthe, Volker Dierschke, Sven Adler

Zusammenfassung

Um das Ausmaß und die Quellen kleinskaliger zeitlicher Variabilität von Seevogelverteilungsmustern in der Deutschen Bucht, südöstliche Nordsee, zu erfassen, wurde eine Reihe von Wiederholungs-Surveys durchgeführt. Es wurden sowohl Flugsurveys als auch schiffsbasierte Erfassungen durchgeführt, um das Ausmaß der Variabilität von Verteilungsmustern auf See zu studieren und den Einfluss von Umweltparametern, die diese Variabilität hervorrufen und steuern, abzuschätzen. Signifikante Unterschiede in der Verteilung und der Abundanz von Seevögeln wurden auf der Ebene einzelner Tage und geringerer Zeitabschnitte nachgewiesen. Zwischenartliche Unterschiede in der Stabilität von Verteilungsmustern wurden auf Unterschiede im Nahrungssuchverhalten zurückgeführt. Die Abundanz, die Verteilung und das Verhalten der Vögel wurde signifikant von einer Reihe von Umweltparametern beeinflusst, z.B. von hydrographischen Fronten, der Windgeschwindigkeit, dem Gezeitenzyklus und der Tageszeit. Diese Faktoren variieren in periodischen oder stochastischen Intervallen und können somit vorhersagbare und nicht vorhersagbare Variationen von Seevogelverteilungsmustern hervorrufen. Die Kenntnis der Ursachen von Variabilität ist somit unerlässlich, um frühere und zukünftige Studien zur Seevogelverteilung zu bewerten, um Monitoringprogramme zur Erfassung von Veränderungen in der Verteilung und Bestandsgröße von Seevögeln zu entwickeln und um Auswirkungen anthropogener Eingriffe zu beurteilen.

Abstract

A series of replicate surveys was conducted to assess extent and sources of small-scale temporal variability in seabird distribution patterns in the German Bight, south-eastern North Sea. Both aerial and ship-based surveys were applied to focus on the extent of variability in distribution patterns and on the influence of environmental parameters that evoke and shape such variability. Significant changes in distribution and abundance of seabirds were recorded at the order of days and smaller time spans. Interspecific differences in stability of distribution patterns were assumed to result from differences in foraging behaviour. Abundance, distribution and behaviour of birds were significantly influenced by a variety of environmental parameters, such as hydrographic fronts, wind speed, tidal stage and time of day. These environmental factors vary at periodic or stochastic intervals and thus have the potential to evoke predictable and unpredictable variations in seabird distribution patterns. Knowledge of sources of variability is essential for the evaluation of results of past and future studies on seabird distribution patterns, the design of monitoring programmes focusing on changes in distribution and numbers, and the evaluation of consequences of anthropogenic impacts.

Introduction

Seabirds are generally not dispersed uniformly at sea but aggregate at different scales due to individual associations and spatially varying availability of resources (Hunt & Schneider 1987, Brown 1988). In addition to this spatial variation, seabird distribution patterns also undergo temporal variation, mostly in response to temporally varying availability of resources (Hunt & Schneider 1987). Temporal variability in distribution patterns is here above all defined as changes in spatial patterns, e.g. shifts in concentration areas. This temporal variability of spatial aspects is further supplemented by variability in overall density, i.e. changes in the total numbers caused by individuals emigrating or immigrating into the study area. Like spatial variability, temporal variability occurs at different scales. Thus, seabird distribution patterns change at the order of decades and years (Final EU Report BECAUSE, in prep.), at the order of seasons and at smaller scales such as days. In the south-eastern North Sea extensive research has focused on seasonal dynamics of seabird distribution patterns (Garthe et al. 2004). At present, mean distribution patterns of all major seabird species can be depicted for each season. However, small- and meso-scale variability can be substantial, thus constraining explanatory power of mean distribution patterns. To our knowledge, an analysis of extent and sources of small-scale variability of distribution patterns was seldom the focus of past seabird studies worldwide (but see Becker et al. 1997, Speckman et al. 2000). Variability of seabird distribution patterns complicates conservation and management efforts as it constrains

(a) the ecological valuation of specific areas,
(b) the calculation of total numbers, and
(c) the assessment of population trends or shifts in distribution.

However, understanding sources of variability enables us to evaluate results of past and future studies on seabird distribution patterns, to design monitoring programmes focusing on changes in distribution and numbers, and to evaluate consequences of anthropogenic impacts.

In the south-eastern North Sea, plans for offshore wind farms necessitated an ecological accompanying research evaluating the construction, maintenance and operation of wind farms with respect to impacts on seabirds among others to enable an ecologically sound and sustainable utilisation of offshore wind energy (Kellermann et al. 2006). In addition to an assessment of general distribution patterns, a description of small- to meso-scale variability within areas of concern was required to complete mean distribution patterns with respect to the ecological valuation of specific areas.

For a number of reasons, the south-eastern North Sea represents both an exceptional and highly suited study area with respect to research on small-scale variability of seabird distribution patterns. First, its hydrography, which forms main habitat structures influencing seabird distribution (Garthe 1997, Skov & Prins 2001, Markones 2007), is characterised by high small- to meso-scale spatial and temporal variability (Becker & Prahm-Rodewald 1980, Dippner 1993). Second, its seabird community mostly comprises gulls (both truly marine species such as Black-legged Kittiwakes *Rissa tridactyla* and coastal species such as *Larus* gulls) and terns, highly mobile opportunistic foragers. Gulls and terns predominantly spend their time flying and thus change position rapidly compared to species swimming most of their time. Moreover, gulls in particular use a variety of different food sources, ranging from fisheries discards and offal to pelagic fish, crustaceans and benthic invertebrates of the Wadden Sea mud flats, and may include even terrestrial organisms from agriculturally dominated

habitats. The different food sources are characterised by varying availability, changing either at periodic intervals, such as benthos availability influenced by tidal rhythms or irregular intervals such as fisheries discards. Due to their high mobility, gulls are able to respond to changing prey availability within short time spans and thus are able to exploit the different habitats effectively. It is assumed that gulls exhibit higher levels of variability in distribution patterns than e.g. Common Guillemots *Uria aalge*, which spend most of their time swimming (Markones 2007).
To analyse extent and sources of variability in seabird distribution patterns, we applied a series of replicate surveys. Emphasis was laid on aerial surveys as they offer the opportunity of sampling large areas within short periods of time. Aerial surveys enabled Nettleship & Gaston (1978) to describe substantial changes in distribution of foraging seabirds within a few days. These shifts in distribution would most probably not have been recognised in the course of ship-based studies as they are progressing at the same temporal scale as potential movements of seabirds (following Brown 1980). However, during ship-based studies a variety of parameters, e.g. with respect to hydrography and food availability, can be sampled simultaneously along with seabird counts. These additional data represent the possibility to gain insight into sources of variability by describing factors influencing occurrence and thus variability of seabird distribution. We consequently applied aerial surveys to assess the extent of variability in large-scale distribution patterns and carried out ship-based studies to focus on the influence of various environmental parameters.

Thus, the present study was set to meet the following objectives: We wanted to

- test the stability of seabird distribution patterns and investigate whether temporal variability is a major characteristic of seabird distribution patterns in the study area
- test for temporal variability at different scales (i.e. at the order of weeks, days and time spans shorter than days)
- evaluate whether some species exhibited higher temporal variability in distribution patterns than others and attempt to explain possible differences
- identify parameters evoking and shaping variability in the study area

Material & Methods

In the German Bight, seabird distribution patterns were recorded by aerial and ship-based surveys. Ship-based surveys followed the internationally standardised "Seabirds at Sea" method described by Tasker et al. (1984), Webb & Durinck (1992) and Garthe et al. (2002). The occurrence of birds was recorded within a 300m wide transect running parallel to the keel line of the observation vessel. Swimming birds were recorded continuously whereas flying individuals were counted following the "snapshot" method (Tasker et al. 1984) to correct for overestimation of particularly mobile species. Densities of all species were obtained by correcting for undetected birds away from the centreline applying the distance-sampling methodology developed by Buckland et al. (1993) and by using functions available in the DISTANCE software package (Laake et al. 1993). This method implied the multiplication of numbers of swimming birds within the observation transect by species-specific correction factors adopted from Garthe et al. (2007). Aerial surveys followed standardised methods described by Kahlert et al. (2000) and Diederichs et al. (2002). Surveys were performed from a high-winged twin-engine Partenavia P-68 with bubble windows at a flight altitude of

78m (250 feet) and a cruising speed of approximately 185km/h (100 knots). The occurrence of birds was recorded to the second within 397m wide transects running parallel to the flight route of the observation platform. Under good observation conditions, both sides of the flight route could be covered, thus resulting in a survey transect width of 794m. Aerial surveys were only carried out during low wind conditions (Table 1) to obtain good observation conditions.

Table 1: Dates of aerial surveys during the breeding season 2006 and respective breeding stages of the two most common breeding species on Helgoland. Classification of breeding stages follows Dierschke et al. (2004), Grunsky-Schöneberg (1998), OAG Helgoland and personal observations.

	7 June	9 June	19 June	28 June	12 July	13 July	12 Aug
Wind	NW 2-3	N 1-2	S-SW 1-3	NW 3-4	S 1-2	SW-W 1-3	N 1-3
Black-legged Kittiwake	Late incubation period		Hatching / Early chick-rearing	Chick-rearing	Late chick-rearing / Fledging		Colony deserted
Common Guillemot	Chick-rearing		Late chick-rearing / Chicks leave ledge	Colony mostly deserted, begin of moult	Colony mostly deserted, begin of moult		Colony deserted, moult

Estimating temporal variability in animal abundances is constrained by sampling variance, i.e. sampling inexactness and spatial variance (Mönkkönen & Aspi 1997). In the "Seabirds at Sea" programme, sampling inexactness is accounted for by avoidance (intercalibration of observers, snapshot method) and correction / mitigation (Buckland distance correction). Spatial variance results from the fact that the size of the sampling unit is inadequate to depict the distribution pattern of individuals in the sampling area (Mönkkönen & Aspi 1997). However, this problem is most relevant with respect to assessing numbers, i.e. variability in density. Although spatial variability can be confused with temporal variability in certain situations, this aspect is not considered to be relevant in the present analysis.

We carried out several dedicated surveys to assess variability of seabird distribution patterns at different temporal scales. In the breeding and immediate post-breeding period of 2006, aerial surveys were conducted along a specifically designed survey route focusing on the waters around the small offshore island Helgoland. The transect course was designed with regard to the breeding species of the seabird colony on Helgoland and covers in particular the central part of the mean distribution area of Black-legged Kittiwakes during the breeding season (Fig. 1). Distance between single transects was set to 8km. Following the method described above, the entire route was sampled within one survey day. Surveys were carried out at different stages of the breeding period to assess differences in distribution patterns at the order of weeks. In addition, we also repeated surveys twice within one or two days, respectively, to assess small-scale variability at the order of days (Table 1). Distribution patterns of Common Guillemots were derived by combining data on Common Guillemot and "Razormots" (Common Guillemot / Razorbill *Alca torda*). This method was considered appropriate as numbers of Razorbills make up less than 1% of numbers of Common Guillemots in summer in the German Bight (Garthe et al. 2007). We tested differences in distribution of Kittiwakes and Common Guillemots between single surveys by modelling the occurrence of the species using the observation position. Thus, a spline of the latitude and longitude values formed the predictor variable in a Generalized Additive Model (GAM, Hastie & Tibshirani 1990, Wood 2006) using the quasi-Poisson function and the MGCV package (Wood 2000) in R 2.5.0 (http://www.r-project.org/). We compared models that were based on data of two survey days with models that were fitted using the data of one of the survey days by applying the ANOVA function (Wood 2006). Thus, we tested differences in the distribution of Kittiwakes between the survey days of 9 June, 19 June, 28 June, 12 July and 12 August to assess variability at the order of weeks. Bonferroni correction was applied to account for multiple testing. In addition, we tested differences in distribution patterns of Kittiwakes and Guillemots between 7 and 9 June, and 12 and 13 July, respectively, to get an indication of variability at the order of days.

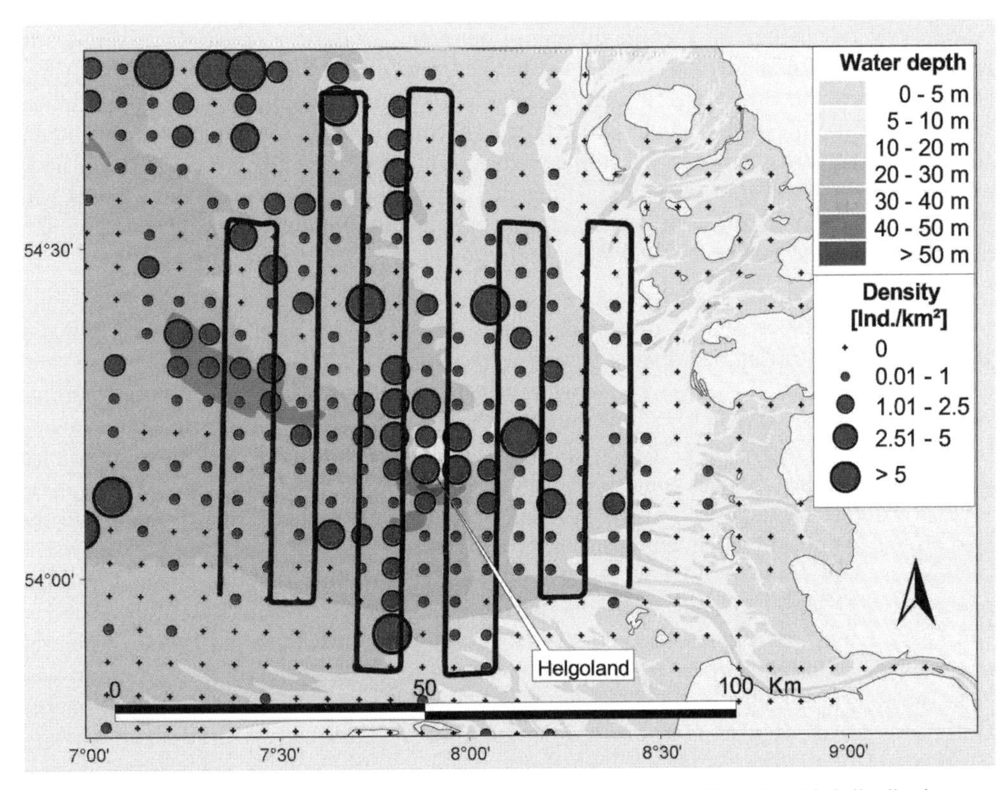

Figure 1: Transect lines of aerial surveys for studying temporal variability of seabird distribution patterns. Additionally depicted is the mean distribution pattern of Kittiwakes during the breeding period (May-July, data source: German Seabirds at Sea database 5.07, comprising ship-based surveys from 1990-2006). Total length of survey transects ~ 840km.

A comparison of the same survey route was also carried out during a ship-based survey in the vicinity of Helgoland (54°11'N, 7°55'E) on 3 and 4 May 2006. On both days, we recorded Kittiwake occurrence on a transect in north-westerly direction from the breeding colony. Sea surface hydrography (temperature and salinity) was assessed simultaneously along with bird counts using Ferrybox instrumentation (http://www.ferrybox.org/, Petersen et al. 2005). Additionally, bird counts were interrupted every 3 nautical miles to record vertical profiles of temperature and salinity continuously from sea surface to sea bottom using a CTD cast. Stations were sampled at the same positions on both days. However, due to unfavourable weather conditions, two stations could not be sampled on 4 May, resulting in a total of 20 stations on 3 May and 18 stations on 4 May. Differences of Kittiwake distribution between the two survey days were tested by use of GAMs following the method described above for aerial surveys. To correlate Kittiwake distribution with hydrography, we divided the seabird survey route into transect segments preceding and following a respective hydrographic station to obtain sampling units that were representative of the Kittiwake occurrence at the station or in its immediate vicinity. Segments were confined to a maximum of 12min sampling time following or preceding the hydrographic station. Thus, bird density was calculated for counting intervals covering a sampling area from 0.94km² to 2.18km² around (i.e. preceding or following) the corresponding hydrographic station. We obtained indicators of frontal occurrence by calculating the average difference between consecutive sea surface salinity and temperature measurements per minute for the respective sampling unit around the hydrographic station. The difference between measurements at the surface and the seabed acted as indication of stratification. In addition to these indicators of fronts (Tfront, Sfront) and stratification (Tdiff, Sdiff), we used sea surface temperature (SST) and sea surface salinity (SSS) as predictor variables in a GAM modelling the distribution of Kittiwakes. The model was fitted using five knots for each parameter and by applying the quasi-Poisson function. The model was selected using backward selection (applying the ANOVA function in R).

Small-scale temporal variability of seabird distribution was additionally studied during ship-based studies in the coastal area between Büsum (54°08'N, 8°52'E) and Helgoland (Fig. 2). Bird counts were carried out along a regular ferry route on ten consecutive days in late summer 2003 from 12 to 21 August. The ferry travelled from Büsum to Helgoland in the morning between 7:30h and 10:15h UTC and returned from Helgoland to Büsum in the afternoon between 14:15h and 17:00h UTC. On 15 August, the outbound survey had to be interrupted due to unfavourable weather conditions. The survey transect presented a cross-section over several different habitats such as the Wadden Sea region, the adjacent Elbe plume and the area around Helgoland which represents a more marine habitat. The route was divided into 4 sections, each comprising 16km of transect length. Thus, seabird distribution could be compared between days, different times of day (morning vs. afternoon) and count sections. Dominant seabird species recorded during the survey between Büsum and Helgoland on the ten ship-based survey days in August 2003 consisted of six gull species and three tern species: Black-headed Gull *Larus ridibundus*, Common Gull *Larus canus*, Lesser Black-backed Gull *Larus fuscus*, Herring Gull *Larus argentatus*, Great Black-backed Gull *Larus marinus*, Black-legged Kittiwake, Sandwich Tern *Sterna sandvicensis*, Common Tern *Sterna hirundo* and Arctic Tern *Sterna paradisaea*. The latter two species were grouped as "Commic Terns" due to identification difficulties.

Differences in daily numbers of each species (group) were tested separately for different count sections and times of day. We applied a Pearson correlation with a 1-day time lag, i.e. tested for correlation of seabird occurrence on a given day with numbers of the following day. Due to missing data (see above) 15 August was ignored in the analysis and 16 August treated as successor of 14 August. Tests were only carried out for species exhibiting at least one positive occurrence (>0) in more than one of nine data pairs. Furthermore, we tested the influence of several environmental parameters on the occurrence of the four most common species, Black-headed Gull, Common Gull, Lesser Black-backed Gull and Herring Gull, during the nine complete survey days.

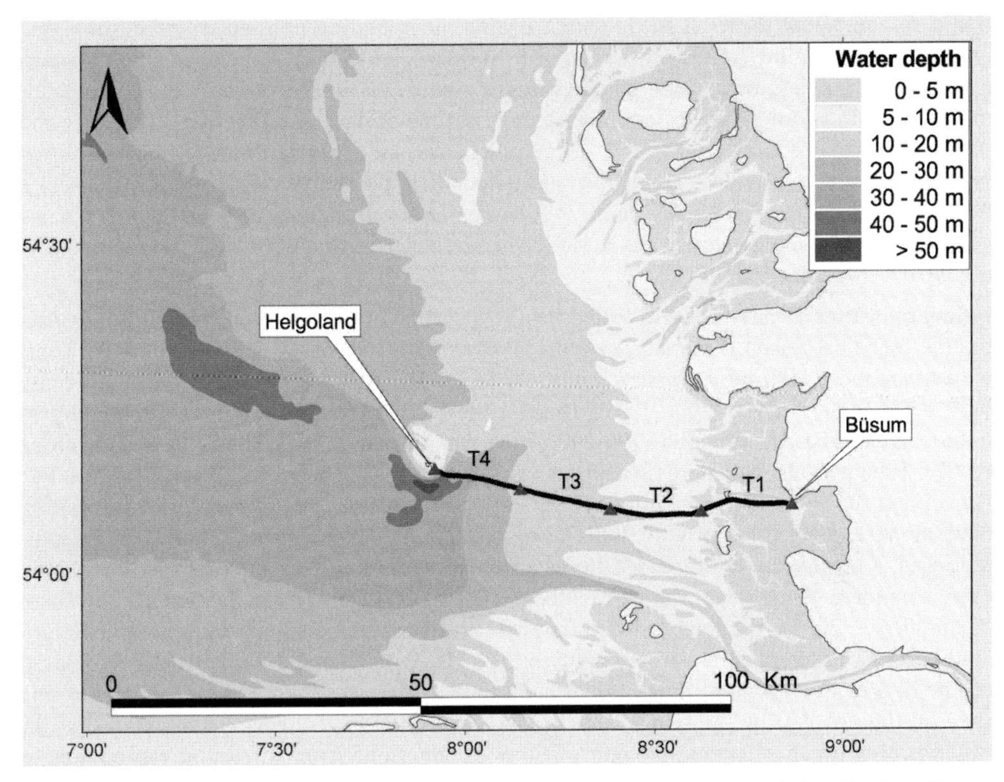

Figure 2: Survey route of ship-based studies between Helgoland and Büsum and classification of count sections T1-4.

Trawler abundance was recorded simultaneously with bird counts to investigate the influence of fishing activity on the occurrence of the four species known to feed regularly on discard and offal. Data on wind speed were obtained from the meteorological station at Research and Technology Centre (RTC) in Büsum. We incorporated maximum wind speed in the analyses as we expected a stronger influence of strong gusts compared to mean wind speed on gull mobility and thus gull occurrence. According to tide forecasts of the Federal Maritime and Hydrographic Agency (BSH), the tidal stage of each count section was classified separately according to day or time of day as either flood tide or ebb tide. Analyses comprised data of count sections 1 and 2 only because gulls predominantly occurred in this area (see Fig. 9). Moreover, trawlers were exclusively recorded in count section 1 and 2. We first tested influence of time of day, count section (= transect), tidal stage and maximum wind speed on trawler abundance, applying a Generalized Linear Mixed Model (GLMM, e.g. Faraway 2006) in R using the library lme4 (Bates & Sarkar 2007). The model was set as follows: response variable = trawler number, predictor = time of day + transect + tide + wind, random effect = day, family = quasi-Poisson. The influence of time of day, count section, tidal stage and maximum wind speed on gull occurrence was tested independently from trawler abundance separately for each species applying a GLMM of the following kind: response variable = bird number, predictor = time of day + transect + tide + wind, random effect = trawler number, family = quasi-Poisson. Influence of trawler abundance on numbers of gulls was tested by comparing these GLMMs with GLMMs incorporating no random effect (only one group) by applying the ANOVA function in R. Influence of trawler abundance on gull occurrence was regarded to be considerable in cases of significant differences between corresponding models. In addition, we tested the influence of time of day, count section, tidal stage and maximum wind speed on resting and scavenging behaviour applying a GLMM of the following kind: response variable = resting / vessel association, predictor = time of day + transect + tide + wind, random effect = day, family = binomial. For the analysis of behaviour, we incorporated data from all ten survey days. Models were selected using backward selection (applying the ANOVA function in R).

Influence of time of day was also tested for the distribution of Black-legged Kittiwakes during the breeding season (May-July) within the area 53.35° to 55.2° N and 5° to 9° E. Thus, we derived mean distribution patterns of this species for 6 different time periods, 2:30–5:59, 6:00–8:59, 9:00–11:59, 12:00–14:59, 15:00–17:59 and 18:00–20:30 h UTC, according to long-term ship-based surveys (1990-2006, data source: German Seabirds at Sea database v5.07). Time periods were hereafter termed as 2-5, 6-8, 9-11, 12-14, 15-17 and 18-20h UTC. No surveys took place before 2:30h UTC and after 20:30h UTC due to darkness. Distribution patterns were mapped in ArcView GIS 3.2 by calculating densities per 3' latitude x 5' longitude grid. Kittiwakes were recorded in distances up to 200km from the breeding site. However, maximum foraging ranges of breeding Kittiwakes in the North Sea amount to 80km (Camphuysen 2005). We thus tested differences in distribution within 80km around Helgoland between different times of day applying GAMs. We compared models that were based on the data of two different times of day with models that were fitted using the data of one specific time of day by applying the ANOVA function (Wood 2006). Thus, we tested for differences between each possible pair of times of day. Bonferroni correction was applied to account for multiple testing.

We then analysed the occurrence of breeding Kittiwakes within different distance classes from the breeding colony on Helgoland. We considered individuals occurring in distances >80km to be entirely independent of the Helgoland breeding colony and distinguished between distance classes of 0-20km, 20-40km, 40-60km and 60-80km. Mean densities were calculated separately for each class of time of day for the four different distance classes on the basis of grid data described above. The influence of time of day and distance from colony was tested on the basis of original data applying a GAM.

Results

The at-sea distribution of Black-legged Kittiwakes around the colony on Helgoland differed substantially between the four different survey days that took place in different stages of the breeding season of 2006. High numbers of Kittiwakes were always recorded in the vicinity of the breeding colony on Helgoland. On 9 June and 12 July, distinct concentrations were additionally observed along the 20m depth contour in the north-west of the study area. High densities were also recorded in the Elbe outflow region east of Helgoland on 12 July. This area along the coast held the majority of Kittiwakes on 19 June. In contrast, it was completely deserted in the post-breeding period on 12 August, as Kittiwakes concentrated in the vicinity of Helgoland and further west and south-west (Fig. 3).

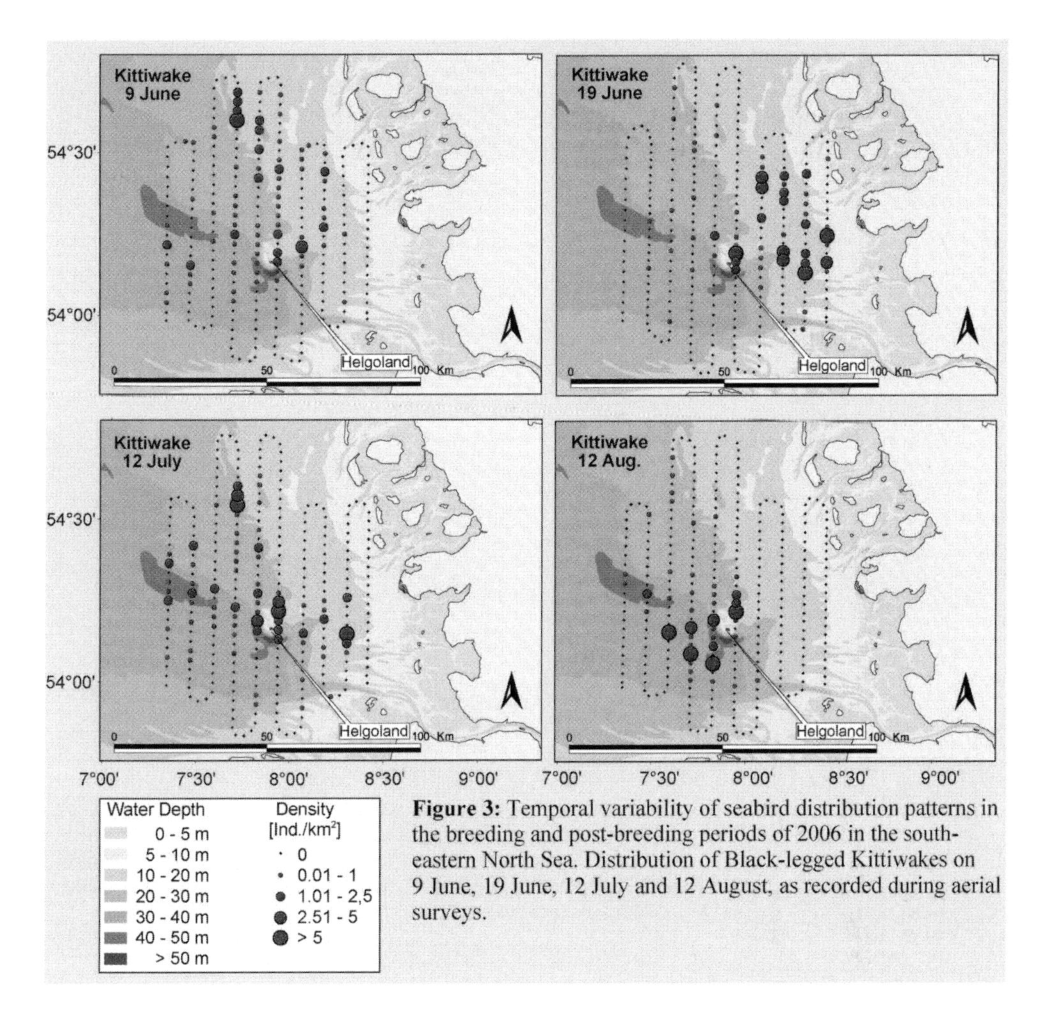

Figure 3: Temporal variability of seabird distribution patterns in the breeding and post-breeding periods of 2006 in the south-eastern North Sea. Distribution of Black-legged Kittiwakes on 9 June, 19 June, 12 July and 12 August, as recorded during aerial surveys.

Distribution patterns differed significantly between single surveys. Only the distribution patterns of 28 June (not depicted) and 12 July were not significantly different, yet both surveys differed significantly from the rest of the days. Significant differences were also detected between consecutive days (Fig. 4). The distribution of Common Guillemots did not differ between 7 and 9 June ($p=0.4$) and was relatively similar between the surveys on the 12 and 13 July 2006 ($p=0.06$, Fig. 4 bottom left & bottom right). In contrast, the distribution of Kittiwakes differed significantly between 7 and 9 June ($p=0.0002$) and 12 and 13 July ($p=0.02$), showing e.g. a concentration area along the 20 m depth contour to the north-west of Helgoland on 12 July that had disappeared the next day (compare Fig. 4 top left & top right).

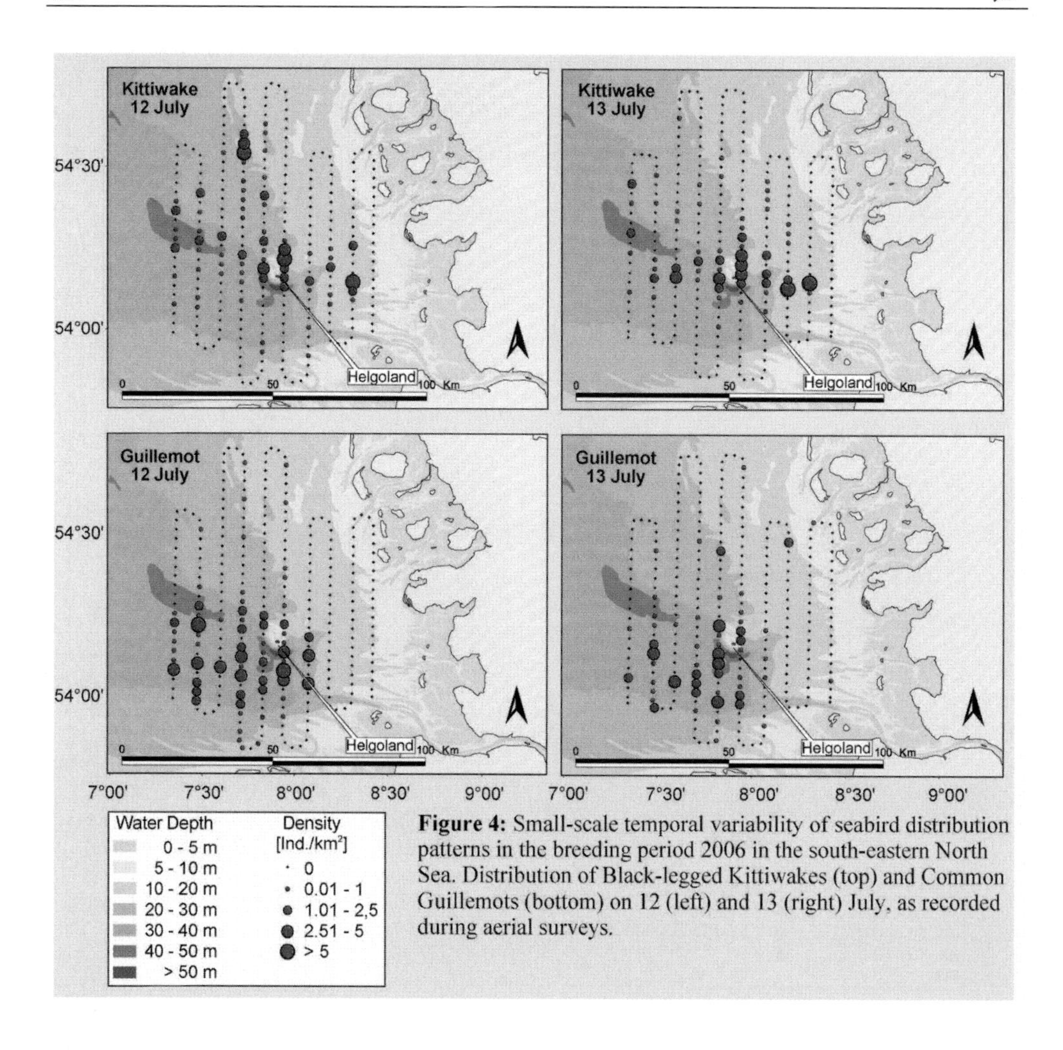

Figure 4: Small-scale temporal variability of seabird distribution patterns in the breeding period 2006 in the south-eastern North Sea. Distribution of Black-legged Kittiwakes (top) and Common Guillemots (bottom) on 12 (left) and 13 (right) July, as recorded during aerial surveys.

Significant differences in Kittiwake distribution were also recorded between 3 and 4 May 2006 during ship-based surveys ($F=8.387$, $p<0.0001$). On both days, high numbers were recorded in the vicinity of the colony. The second concentration area at sea, however, shifted further to the north-west from 3 to 4 May (Fig. 5 A and Fig. 5 B). The occurrence of Kittiwakes was best explained by a GAM comprising sea surface salinity (SSS), and indicators of salinity fronts (Sfront) and temperature stratification (Tdiff). All parameters influenced Kittiwake occurrence significantly (variance of the intercept: -2.586, $p=0.02$; SSS: $F=5.091$, $p=0.031$; Sfront: $F=5.156$, $p=0.003$; Tdiff: $F=5.030$, $p=0.003$).

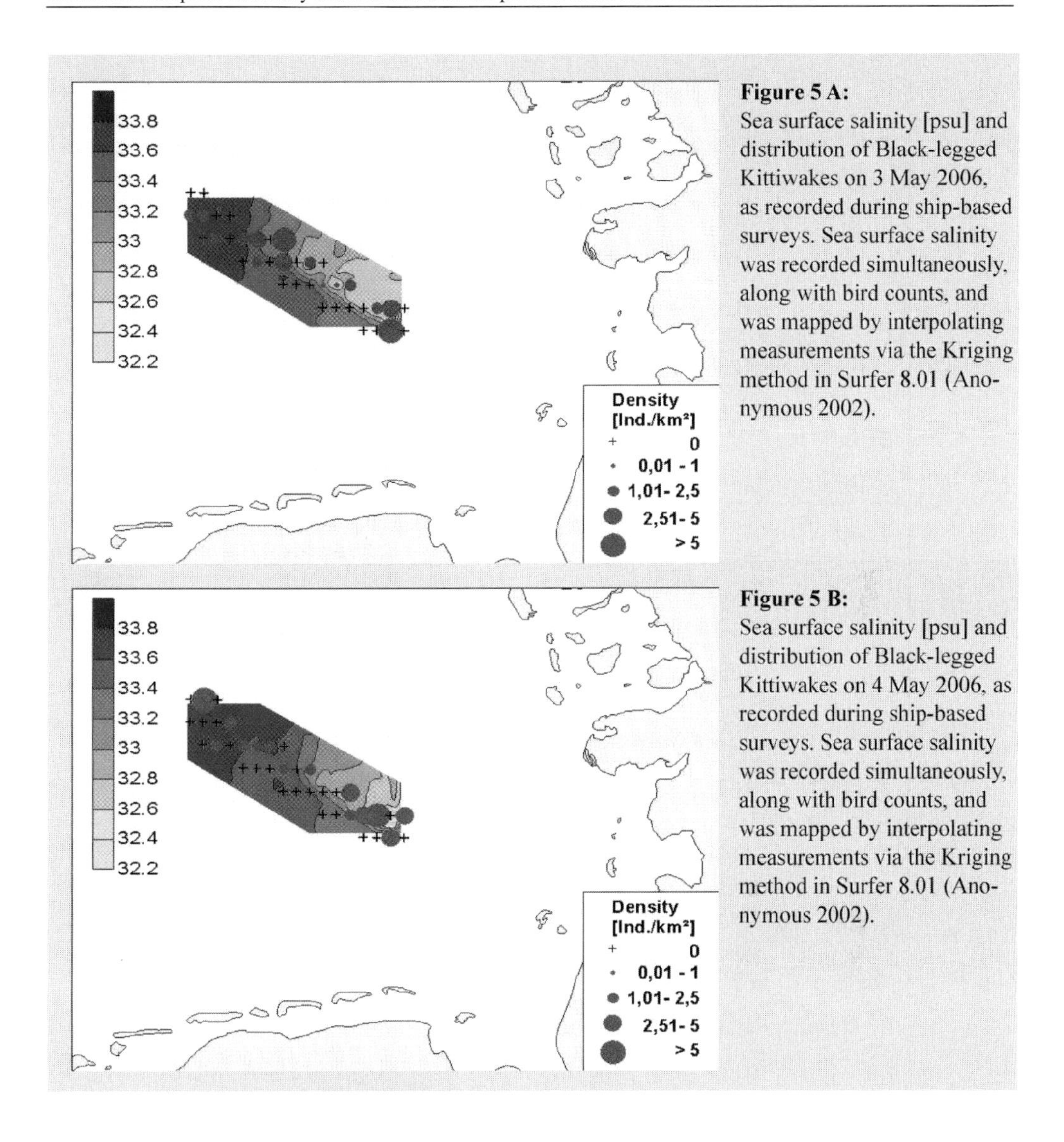

Figure 5 A:
Sea surface salinity [psu] and distribution of Black-legged Kittiwakes on 3 May 2006, as recorded during ship-based surveys. Sea surface salinity was recorded simultaneously, along with bird counts, and was mapped by interpolating measurements via the Kriging method in Surfer 8.01 (Anonymous 2002).

Figure 5 B:
Sea surface salinity [psu] and distribution of Black-legged Kittiwakes on 4 May 2006, as recorded during ship-based surveys. Sea surface salinity was recorded simultaneously, along with bird counts, and was mapped by interpolating measurements via the Kriging method in Surfer 8.01 (Anonymous 2002).

High intensity of salinity fronts had a distinct positive effect on Kittiwake distribution (Fig. 6). The model explained 65.8% of the total variance of the data. However, the model was based on 38 data points only and removing the data point of maximum density resulted in less significant effects (Fig. 6 & 7) and a lower value of 52.4% variance explained by the model.

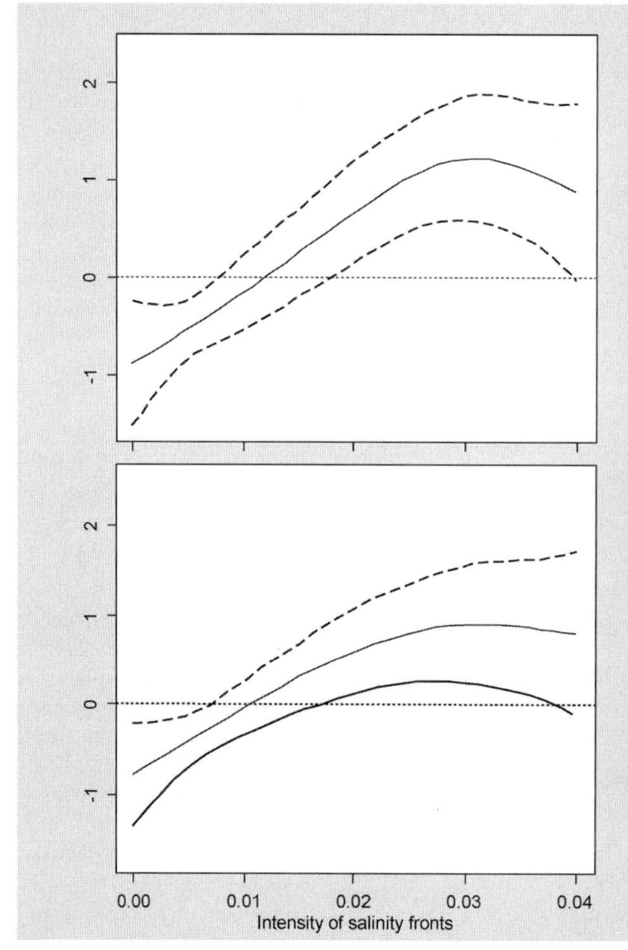

Figure 6:
GAM smoothing curves fitted to partial effects of explanatory variables on density of Black-legged Kittiwakes uring ship-based studies on 3 and 4 May. Density is represented as a function of intensity of salinity fronts depicted for a GAM based on complete data of both study days (top; F=5.156, p=0.003), and for a GAM based on data lacking the data point of maximum density (see Fig. 7; F=2.989, p=0.03; right).
Dashed lines represent 95% confidence intervals around the main effects. High intensity of salinity fronts had a positive effect on Kittiwake distribution. Frontal intensity represents an indication of gradient strength and frequency of frontal incidents (see Methods).

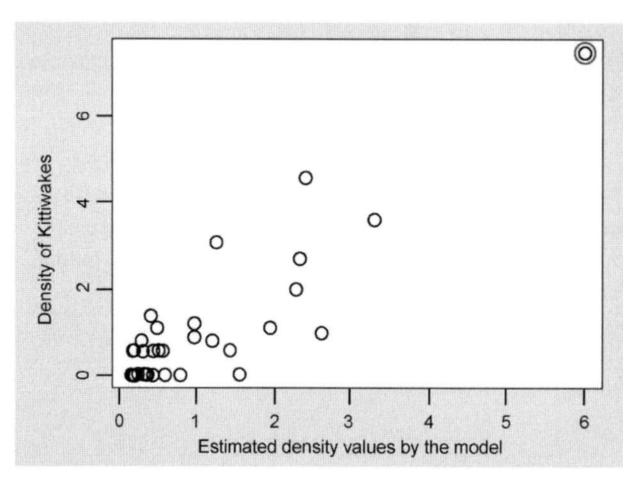

Figure 7:
Estimated density by GAM versus measured density. High correlation is caused by the data point of maximum density encircled in red.

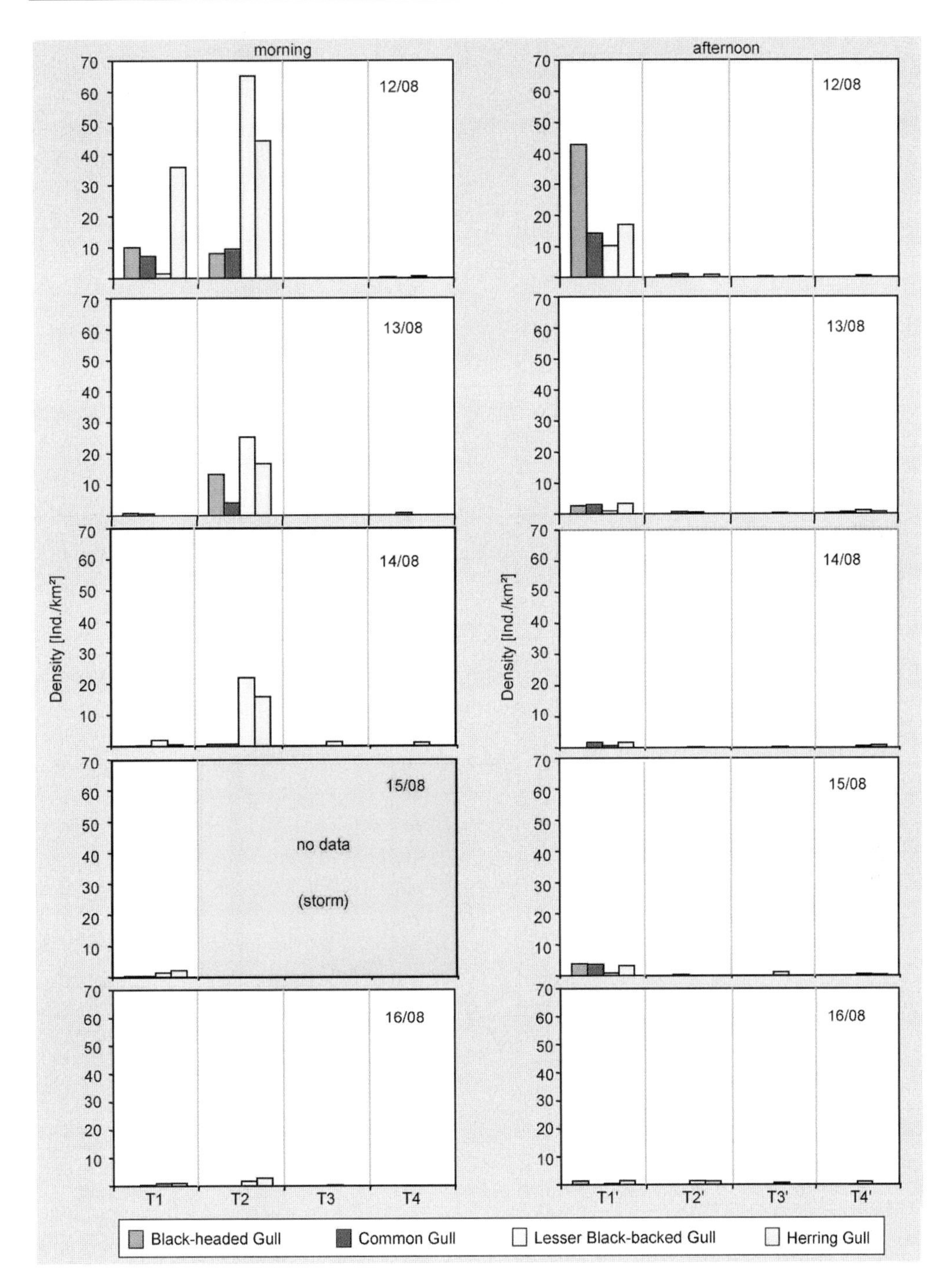

Figure 8 A: Variability of seabird occurrence during repeated surveys in the study area between Büsum and Helgoland in August 2003 (12-16 August).

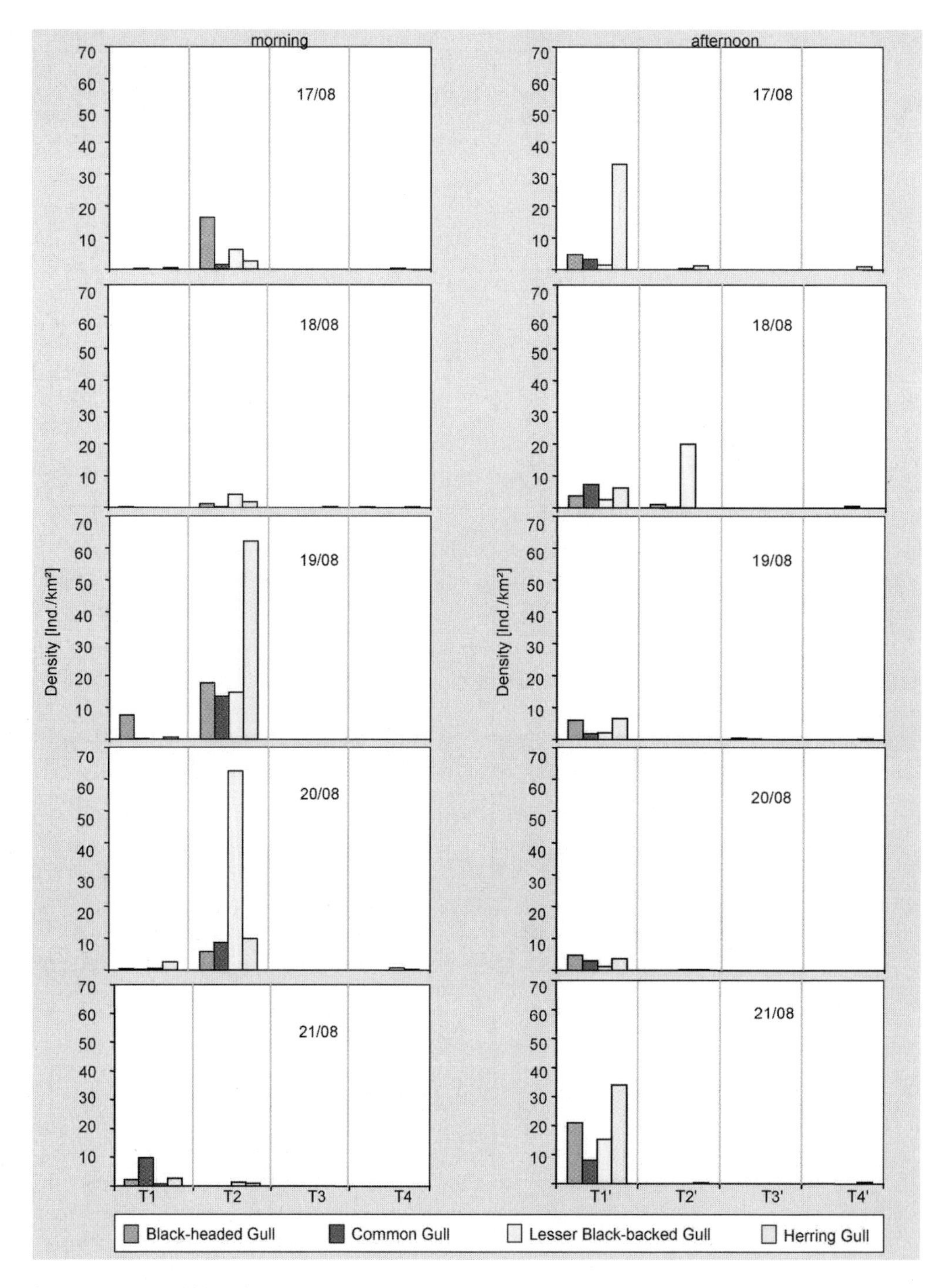

Figure 8 B: Variability of seabird occurrence during repeated surveys in the study area between Büsum and Helgoland in August 2003 (17-21 August).

The occurrence of seabird species between Büsum and Helgoland on the ten ship-based survey days in August 2003 was characterised by high variability. Numbers of different species peaked at different times, and patterns of occurrence differed between the morning and afternoon trip and between consecutive days (Fig. 8 A and Fig. 8 B). Accordingly, analyses revealed no correlation of the occurrence of all eight gull and tern species (groups) between consecutive days: Only two of 55 Pearson correlation tests produced a significant result, constituting a proportion of less than 2%. Thus, the proportion of significant results falls within the expected range (<0.05%) of a dataset holding no effect. In addition, the mean correlation coefficient (mean rho) equals approximately 0, implying no trend. However, numbers did not vary randomly but were correlated to a variety of factors. Thus, trawler abundance correlated significantly positively with the occurrence of Black-headed, Common, Lesser Black-backed and Herring Gulls. Trawler occurrence itself was significantly influenced by time of day, count section and wind, but not by tidal stage. Numbers of trawlers were higher in the second count section further offshore and decreased with time of day and with increasing wind speed. Thus, the influence of the factors time of day, count section, tidal stage and wind had to be tested independently from influence of trawler abundance. All four factors had a significant influence on the occurrence of all four species (significant at 0.05 level, n observations = 36, n groups of trawler abundance = 8). Numbers of all four gull species decreased with increasing wind velocity and were higher during flood tide compared to ebb tide. While numbers of Black-headed, Common and Herring Gulls correlated positively with time of day and negatively with distance to coast, Lesser Black-backed Gulls showed an inverse relationship, with higher numbers in the morning and in the second count section further offshore.

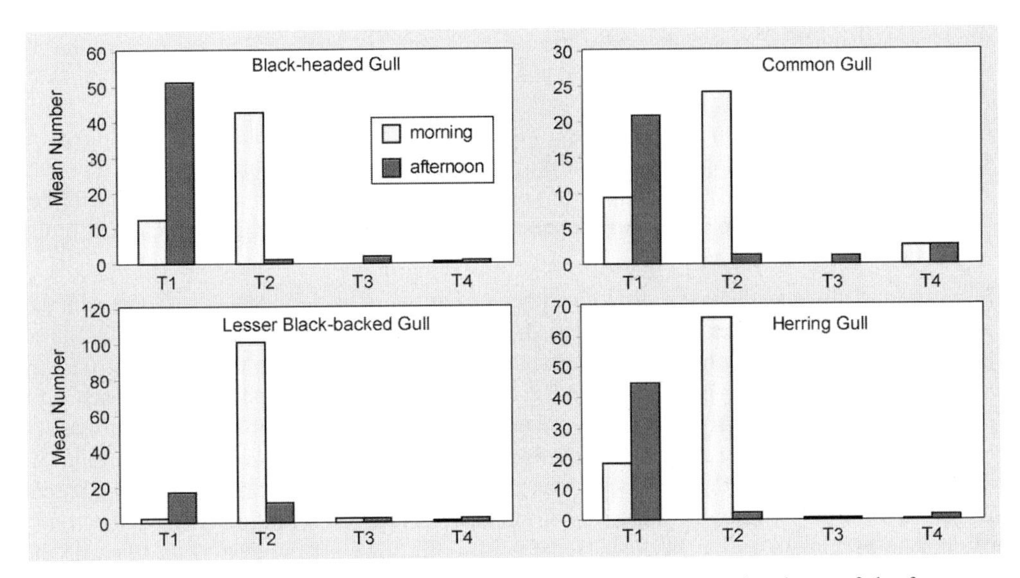

Figure 9: Spatially and temporally differing area usage by gulls. Mean abundance of the four most common gulls in count sections T1-T4 during ship-based surveys between Büsum and Helgoland in August 2003.

An overview of total numbers of gulls in each count section during the different times of day depicted striking spatio-temporal differences in area usage (Fig. 9). All species occurred predominantly in the first and second count sections in the Wadden Sea region. In the morning, highest gull numbers occurred in count section 2, whereas in the afternoon most gulls occurred in count section 1 closer to the coast. Trawler numbers were higher in the morning in count section 2 but did not differ between times of day in count section 1. In addition to the observed differences in numbers, all four gull species also showed differences in behaviour between times of day (Fig. 10).

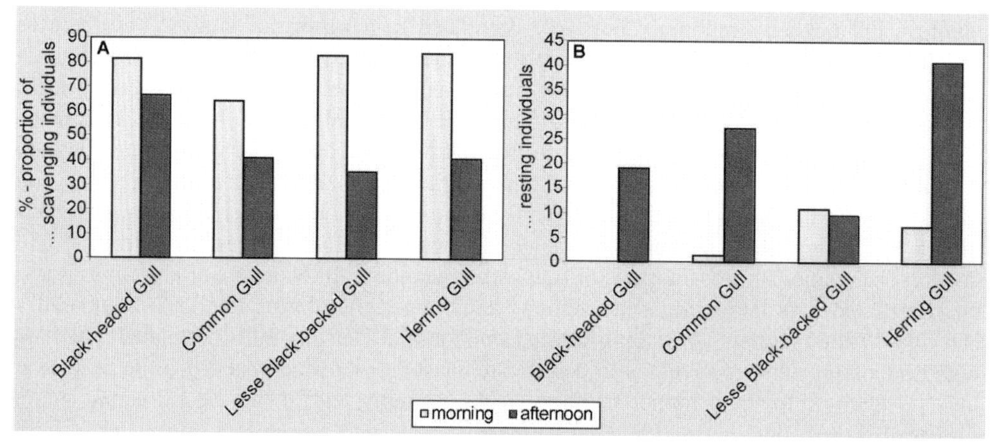

Figure 10: Differences between morning and afternoon in scavenging (A) and resting (B) behaviour of Black-headed Gull, Common Gull, Lesser Black-backed Gull and Herring Gull observed during repeated surveys between Büsum and Helgoland in August 2003.

Levels of association with fishing vessels were generally higher in the morning, higher in section 2 and higher during ebb tide and in periods of low winds. Common Gulls however showed highest values of association with vessels in the afternoon. Resting behaviour correlated significantly positively with high winds, flood tide, count section 2 and increased in the afternoon (Table 2, see page 132).

Significant differences between different times of day were also evident for the distribution of Black-legged Kittiwakes in the German Bight during the breeding season, with low numbers at sea in the early morning and late evening and high numbers in the middle of the day (Figs. 11 & 12, following pages). According to GAMs, the distribution pattern of each time of day differed significantly from every other time of day. Density of birds was influenced significantly negatively by distance from colony ($F=41.69$, $p<0.0001$) and varied with time of day ($F=12.30$, $p<0.0001$, Fig. 12). Being constricted to the ultimate vicinity around the breeding colony in the morning, the main concentration area expands with time of day, apparently reaching a maximum extent between 12 and 14h UTC and contracting again towards the evening (Fig. 11). Accordingly, highest mean densities in the distance classes 40-60km and 60-80km were recorded for the period between 12 and 14h UTC. However, mean density values in the nearest distance class 0-20km were highest throughout the day, reaching values of 1.8 individuals/km^2.

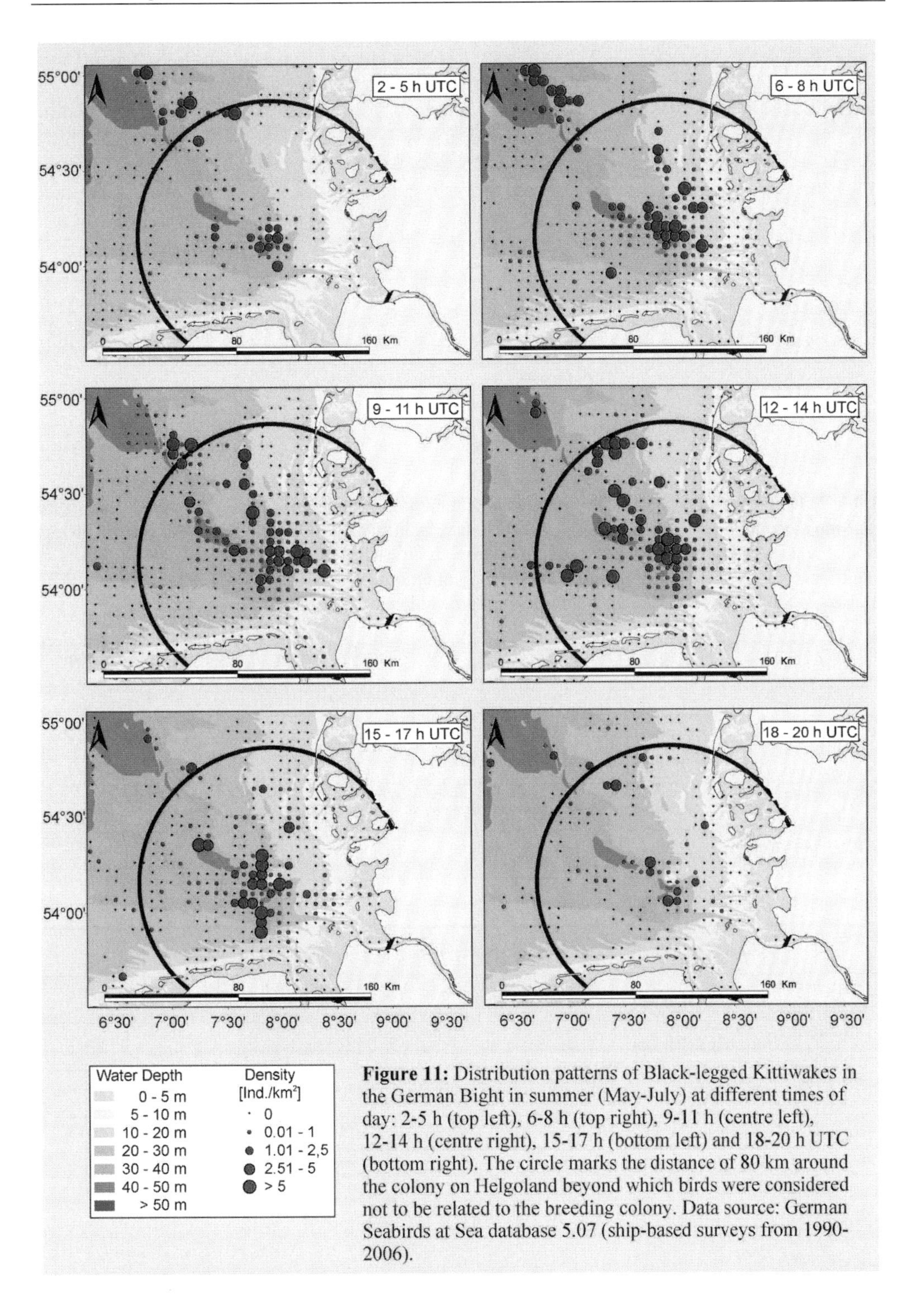

Figure 11: Distribution patterns of Black-legged Kittiwakes in the German Bight in summer (May-July) at different times of day: 2-5 h (top left), 6-8 h (top right), 9-11 h (centre left), 12-14 h (centre right), 15-17 h (bottom left) and 18-20 h UTC (bottom right). The circle marks the distance of 80 km around the colony on Helgoland beyond which birds were considered not to be related to the breeding colony. Data source: German Seabirds at Sea database 5.07 (ship-based surveys from 1990-2006).

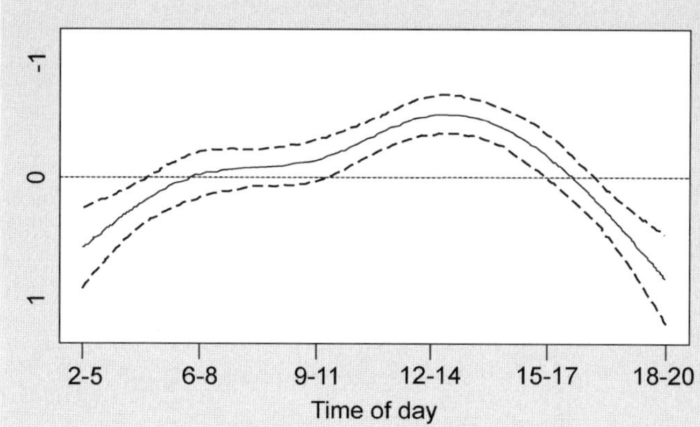

Figure 12: GAM smoothing curve fitted to partial effects of an explanatory variable on den-sity of Black-legged Kittiwakes. Density is represented as a func-tion of time of day. Dashed lines represent 95% confidence intervals around the main effects. Maximum densities were recorded between 12h and 14h UTC. Data source: German Seabirds at Sea database v5.07.

Table 2: Results of GLMMs testing the influence of environmental parameters on the behaviour of the four most common gull species during repeated surveys in the study area between Büsum and Helgoland. Parameters significantly influencing the response variable are listed under predictors (level of significance: 0.05). Dtime = time of day (morning / afternoon), tide = tidal stage (ebb tide / flood tide), transect = count section (T1 / T2), wind = maximum wind velocity recorded in m/s.

Species	Response variable	Predictors	Random effect	No. observations	No. groups / days
All	Vessel association	Dtime+transect+tide+wind	Day	3702	10
Black-headed Gull	Vessel association	Transect+tide+wind	Day	832	10
Common Gull	Vessel association	Dtime+transect+tide+wind	Day	478	10
Lesser black-backed Gull	Vessel association	Dtime+wind	Day	1190	10
Herring Gull	Vessel association	Dtime+transect+wind	Day	1202	10
All	Resting behaviour	Dtime+tide+transect+wind	Day	3702	10
Black-headed Gull	Resting behaviour	Dtime+transect	Day	832	10
Common Gull	Resting behaviour	Dtime+tide	Day	478	10
Lesser black-backed Gull	Resting behaviour	Transect+wind	Day	1190	10
Herring Gull	Resting behaviour	Dtime+transect+tide+wind	Day	1202	10

Discussion

Methodology

To discuss the results presented it is important to evaluate whether we assessed representative data during the analysed surveys. Aerial surveys covered the main distribution area of both Kittiwakes and Guillemots in the breeding season. However, to achieve a survey route that could be covered within one day, it was not possible to extend surveys to the maximum foraging distance of 80km. Instead, maximum distance to the north of the breeding colony reached by aerial surveys amounted to 60km. Aerial surveys were carried out above all between 7 and 13h UTC. This implies that

(a) distribution possibly varied during the survey according to different times of day and
(b) surveys took place at least partly within the period of maximum extension of distribution (according to the above results).

However, individual surveys were always carried out at the same time of the day. Thus, we assume that comparisons between individual surveys were not constrained by the methodology applied. Another critical aspect concerns the spatial effort. We covered approximately 10% of the survey area. However, some sections had to be excluded from analyses due to unfavourable observation conditions. Nevertheless, the overall sampling effort was considered high so that the probability of depicting general distribution patterns correctly was high, although smaller concentration areas may have been overlooked due to transect layout.

Results of correlation of Kittiwake distribution with hydrographic parameters on 3 and 4 May 2006 were based on a limited dataset of two days only in a spatially restricted area. Consequently, variance can be assumed to be low. The GAM explained a relatively high proportion of total variance but confidence intervals are rather high (see e.g. Fig. 6). Removing the data point of maximum density altered results pronouncedly (Figs. 6 & 7). In conclusion, the significant effects presented here should be regarded as tendencies implying potential interrelations. Nevertheless, our results fit earlier studies based on extensive datasets which also report a significant positive effect of fronts on Kittiwake distribution (see below; Markones 2007).

Ship-based surveys between Büsum and Helgoland covered a very limited proportion of the total area only and thus were not suited to depict distribution patterns. However, as the survey transect presented a section through different habitats, general patterns of occurrence can be described, particularly with regard to differences between the different regions. The main focus of this paper, however, was on temporal variability. The studied species, i.e. gulls and terns, can be assumed to exhibit higher temporal variability in their occurrence than other seabird species because they are highly mobile opportunistic foragers. Thus, the results cannot be fully extrapolated to other seabird species which spend most of their time swimming or possess a more restricted prey spectrum. Most gulls also forage in terrestrial habitats and on mud flats of the Wadden Sea region. Thus, not only distribution but also numbers of gulls at sea vary with time. The study took place in the post-breeding season when birds are not bound to the colony anymore but are able to roam freely. Migration has started, possibly causing changes in numbers of birds. Results of our study do not refer to actively migrating individuals but focus on mostly stationary birds foraging and resting in the area. Our study may represent an extreme situation of high variability due to the study period and the species studied. Applying this method, no reliable information can be derived on changes in numbers of studied species due to our low spatial effort. However, we were able to give information on several aspects of variability in seabird distribution patterns. We

(a) assessed a possible order of magnitude of temporal variability in seabird distribution areas,
(b) depicted general shifts in concentration areas, i.e. changes in distribution patterns, with time of day, and
(c) identified environmental parameters causing and shaping variability of seabird occurrence.

Influence of time of day on distribution of Kittiwakes was tested using a rather heterogeneous dataset characterised by efforts varying spatially and temporally (Table 3).

Table 3: Survey effort of long-term ship-based surveys (1990-2006, German Seabirds at Sea database v5.07) within the four different distance classes around the colony on Helgoland at different times of day. Effort is given as the ratio between the total survey area (including repeated surveying of the same plots) and the area size covered by each distance class.

	2-5 h	**6-8 h**	**9-11 h**	**12-14 h**	**15-17 h**	**18-20 h**
0-20 km	0.14	0.41	0.53	0.49	0.16	0.05
20-40 km	0.02	0.19	0.16	0.09	0.10	0.03
40-60 km	0.01	0.08	0.06	0.05	0.06	0.01
60-80 km	0.01	0.02	0.02	0.02	0.02	0.01

Ideally, the same area should have been sampled repeatedly at all different time periods during one single day. However, a realisation of this method is possible neither by ship nor by plane due to the large foraging area of breeding Kittiwakes. On the other hand, bearing the present results on variability of Kittiwake distribution patterns at the order of single days in mind, mean distribution patterns can be assumed to be less influenced by singular events. The dataset may nevertheless produce reasonable results because it is based on a large number of different survey days. Furthermore, we were unable to distinguish between breeding and non-breeding individuals. Distribution patterns, especially in the distant areas, may have been influenced by birds not breeding on Helgoland. However, bias presumably was low as breeding birds make up the majority of numbers of Kittiwakes during the breeding season (following Garthe et al. 2007).

Scales of temporal variability

Distribution patterns of Black-legged Kittiwakes differed significantly in most cases between the different phases of the breeding season, indicating variability of distribution patterns at the order of weeks. Results from repeated specific surveys on the following day emphasised high variability at the order of days. As surveys were carried out on single days, variability at the order of weeks, as found in this study, may represent variability at the order of days. It should be noted that weeks do not represent biologically distinct time periods but may be considered multiples of days. The time period of one day will be reflected in most species' occurrence because the majority of seabirds exhibits diurnal patterns in activity (e.g. Shealer

2002). Variability at a temporal scale finer than one day is caused by factors varying by hours or minutes such as tidal stage, wind speed and direction.

Interspecific differences

The higher variability found in distribution patterns of Black-legged Kittiwakes compared to Common Guillemots may result from the fact that Kittiwakes are strongly associated with short-lived frontal structures (Markones 2007). In contrast, Common Guillemots are not restricted to prey patches at the sea surface but are able to use the whole water column, i.e. a third dimension of their habitat. Regarding only horizontal distribution patterns, Kittiwakes are likely to exhibit a higher variability than Common Guillemots, which may alter their diving behaviour in response to differing food availability. Kittiwakes mostly spend their at-sea time flying (Markones 2007) and cover larger distances during a shorter time span. Common Guillemots spend most of their at-sea time swimming (Markones 2007). Consequently, distribution patterns of Kittiwakes can vary at shorter intervals due to a higher amount of travelling involved.

Influence of environmental parameters

The observed significant variations of seabird distribution patterns at different scales could partly be explained by various environmental factors such as trawler abundance, hydrographic parameters, wind speed, tidal stage and time of day. These factors influence seabird occurrence in different ways. Trawler abundance acts as a measure for direct food availability, with fishing vessels representing an anthropogenic food source. Hydrography and tide may influence abundance, distribution and behaviour of prey species of seabirds in the south-eastern North Sea, such as polychaetes (Esselink & Zwarts 1989), crustaceans (Aagaard et al. 1995) and fish (Valenzuela et al. 1991, Thiel et al. 1995). These factors act as indirect indicators of food availability. The correlation of Kittiwakes with frontal structures in the German Bight has been recorded earlier and is assumed to result from enhanced food availability at fronts due to physical and biological mechanisms (Markones 2007).

Wind speed influences both mobility of birds and food availability. Medium wind speeds may reduce flight costs (e.g. Furness & Bryant 1996) while strong winds constrain flight performance and thus mobility (Woodcock 1940, Manikowski 1971). Moreover, strong winds reduce fishing activity (e.g. see results) and decrease natural food availability (Dunn 1973, Finney et al. 1999). Time of day may reflect food availability, e.g. in case of prey species exhibiting diel vertical migration such as zooplankton and fish (Hays 2003). The link between time of day and seabird occurrence is also influenced by intrinsic factors such as the behaviour of breeding birds to spend the night at the colony. During hours of darkness, Common Guillemots do not forage (Wanless et al. 1988) and Kittiwakes do not fly (Daunt et al. 2002). While resting at sea during the night of the early breeding season (May – mid of June), Kittiwakes spend the night at the colony during chick-rearing (mid of June – mid of July, Glutz von Blotzheim & Bauer 1999). This behaviour is reflected in varying distribution patterns studied at different times of the day during the breeding season (Fig. 11). As the distribution ranges furthest from the colony during mid-day, one may conclude that the furthest areas represent especially favourable foraging grounds that are only within reach of Helgoland breeding birds at this time of day because of the long distance from the breeding colony. However, assuming that individuals leave the colony at dawn and travel at an average flight speed of 47km/h (Pennycuick 1997), the maximum foraging range of 80km can be reached within less than 2 hours time, i.e. at approximately 4h UTC and thus much earlier than observed. A different explanation could lie in prey availability varying with time of day such that birds encounter favourable food availability at noon / early afternoon only. However, this seems unlikely as most prey species of seabirds rather retreat out of reach during daylight (Hays 2003). Large distances from the Helgoland colony reached during 12-14h UTC however coincide with by far the highest numbers of birds at sea. Thus, birds might simply spread over a larger area to avoid competition with conspecifics.

Our results correspond to other studies that found the early afternoon to be one of the periods of high activity in breeding Kittiwakes (Glutz von Blotzheim & Bauer 1999). However, the early morning, which is generally characterised by highest activity, did not stand out in our study. Chick-rearing Kittiwakes equipped with data loggers showed highest foraging activity and highest amounts of time at sea during the morning (~8-10h UTC) and high values of colony attendance in the early afternoon (14-15h UTC, Daunt et al. 2002).

Behaviour of Black-headed, Common, Lesser Black-backed and Herring Gulls differing with time of day can be explained by the fact that seabirds - like most birds - show diurnal activity peaks and display e.g. intensified feeding early in the morning and late in the afternoon (Bezzel & Prinzinger 1990). The fact that gulls occurred closer to shore in the afternoon may result from a movement to roosting aggregations at land, where gulls mostly spend the nights outside the breeding season.

Conclusions

Temporal variability has been shown to be a major characteristic of seabird distribution patterns. Some of the environmental factors influencing variability, such as season, time of day and tide, are of a periodic nature. In contrast, wind speed, hydrography (which is often influenced by wind speed and direction) and trawler abundance are shaped by a variety of different factors and are characterised by a more or less stochastic quality. Thus, seabird distribution patterns undergo predictable as well as unpredictable temporal variations at fine scales. Differences can be significant, requiring the consideration of temporal variability during the design and interpretation of surveys and ecological assessments. Future studies need to proceed in identifying other sources of variability and characterising relationships between seabird occurrence and environmental parameters. Knowledge of these effects can e.g. maximise survey effectiveness with respect to estimating numbers and trends by sampling during periods of lowest variability (Becker et al. 1997, Speckman et al. 2000) and would permit an improved interpretation of survey results.

Acknowledgements

We would like to thank the following individuals and institutions for their invaluable contributions: Aerial and ship-based surveys comprise data collected with financial support of the Federal Ministry for the Environment, Nature Conservation and Nuclear Safety (BMU) and the Federal Agency for Nature Conservation (BfN) as well as the Verein der Freunde und

Förderer der Inselstation der Vogelwarte Helgoland e.V., the Ornithologische Arbeitsgemeinschaft für Schleswig-Holstein & Hamburg e.V. and the State Office for the National Park Schleswig-Holstein Wadden Sea. The analysis was funded in part by the project "Zeitlich-räumliche Variabilität der Seevogel-Vorkommen in der deutschen Nord- und Ostsee und ihre Bewertung hinsichtlich der Offshore-Windenergienutzung" (project 5 of MINOS+, financed by the Federal Environmental Ministry). Kai-Uwe Breuel (Sylt Air), Stefan Hecke and Felix Schmittendorf (FLM Aviation Kiel) steered the plane with great skill during aerial surveys. The shipping company Cassen Eils granted access to the ferry "Atlantis" between Büsum and Helgoland. Volker Dzaak (GKSS Research Centre) coordinated access to RV "Ludwig Prandtl". The crew of RV "Ludwig Prandtl" (GKSS), Helmut Bornhöft and Jan Marx, and the crew of the "Atlantis" provided excellent opportunities to work on board. Martina Gehrung (Institute for Coastal Research / Operational Systems, GKSS) kindly enabled the use of the Ferrybox and kept it running. Gero Bojens and Klaus Vanselow (Applied Physics / Marine Technology, FTZ) provided and maintained the CTD cast. Wind data was obtained from the meteorological station of the FTZ (K. Vanselow). Jana Kotzerka, Moritz Mercker, Britta Meyer, Dennis Münd, Tanja Weichler and many others assisted in bird counts during ship-based surveys. Ommo Hüppop and the team of the Institut für Vogelforschung "Vogelwarte Helgoland" Inselstation provided accommodation on Helgoland during ship-based surveys. Roger Mundry assisted in earlier statistical analyses. Franciscus Colijn gave valuable comments on an earlier draft of the manuscript.

References

Aagaard A, Warman CG & Depledge MH (1995). Tidal and seasonal changes in the temporal and spatial distribution of foraging *Carcinus maenas* in the weakly tidal littoral zone of Kerteminde Fjord, Denmark. Mar. Ecol. Prog. Ser. 122:165-172.

Anonymous (2002). Surfer 8. Contouring and 3D surface mapping for scientists and engineers. Golden Software, Colorado.

Bates D & Sarkar D (2007). Reference manual lme4: Linear mixed-effects models using S4 classes. http://mirrors.dotsrc.org/cran/.

Becker BH, Beissinger SR, Carter HR (1997). At-sea density monitoring of Marbled Murrelets in central California: methodological considerations. Condor 99:743-155.

Becker GA & Prahm-Rodewald G (1980). Fronten im Meer. Salzgehaltsfronten in der Deutschen Bucht. Seewarte 41:12-21.

Bezzel E & Prinzinger R (1990). Ornithologie. Ulmer, Stuttgart.

Brown RGB (1980). Seabirds as marine animals. In: Burger J, Olla B & Winn HE (eds.): Marine birds. Plenum Press, New York, pp. 1-39.

Brown RGB (1988). Zooplankton patchiness and seabird distributions. Acta XIII Int. Ornithol. Congr. 1:1001-1009.

Buckland ST, Anderson DR, Burnham KP & Laake JL (1993). Distance sampling. Estimating abundance of biological populations. Chapman and Hall, London.

Camphuysen CJ, (ed.) (2005). Understanding marine foodweb processes: an ecosystem approach to sustainable sandeel fisheries in the North Sea. IMPRESS Final Report, Royal Netherlands Institute for Sea Research, Texel.

Daunt F, Benvenuti S, Harris MP, Dall'Antonia L, Elston DA, Wanless S (2002). Foraging strategies of the Black-legged Kittiwake *Rissa tridactyla* at a North Sea colony: evidence for a maximum foraging range. Mar. Ecol. Prog. Ser. 245:239-247.

Diederichs A, Nehls G, Pedersen IK (2002). Flugzeugzählungen zur großflächigen Erfassung von Seevögeln und marinen Säugern als Grundlage für Umweltverträglichkeitsstudien im Offshorebereich. Seevögel 23:38-46.

Dierschke V, Garthe S, Markones N (2004). Aktionsradien Helgoländer Dreizehenmöwen *Rissa tridactyla* und Trottellummen *Uria aalge* während der Aufzuchtphase. Vogelwelt 125:11-19.

Dippner JW (1993). A frontal-resolving model for the German Bight. Cont. Shelf Res. 13: 49-66.

Dunn EK (1973). Changes in fishing ability of terns associated with windspeed and sea surface conditions. Nature 244:520-521.

Esselink P & Zwarts L (1989). Seasonal trend in burrow depth and tidal variation in feeding activity of *Nereis diversicolor*. Mar. Ecol. Prog. Ser. 56:243-254.

Faraway JJ (2006). Extending the linear model with R: generalized linear, mixed effects and nonparametric regression models. Chapman and Hall, London.

Finney SK, Wanless S & Harris MP (1999). The effect of weather conditions on the feeding behaviour of a diving bird, the Common Guillemot *Uria aalge*. J. Avian Biol. 30:23-30.

Furness RW & Bryant DM (1996). Effect of wind on field metabolic rates of breeding Northern Fulmars. Ecology 77:1181-1188.

Garthe S (1997). Influence of hydrography, fishing activity, and colony location on summer seabird distribution in the south-eastern North Sea. ICES J. Mar. Sci. 54:566-577.

Garthe S, Hüppop O, Weichler T (2002). Anleitung zur Erfassung von Seevögeln auf See von Schiffen. Seevögel 23:47-55.

Garthe S, Dierschke V, Weichler T, Schwemmer P (2004). Teilprojekt 5 - Rastvogelvorkommen und Offshore-Windkraftnutzung: Analyse des Konfliktpotenzials für die deutsche Nord- und Ostsee. In: Kellermann A et al. (Eds.): Marine Warmblüter in Nord- und Ostsee: Grundlagen zur Bewertung von Windkraftanlagen im Offshore-Bereich. Endbericht.

Garthe S, Sonntag N, Schwemmer P, Dierschke V (2007). Estimation of seabird numbers in the German North Sea throughout the annual cycle and their biogeographic importance. Vogelwelt 128: *in press*.

Glutz von Blotzheim UN & Bauer KM (1999). Handbuch der Vögel Mitteleuropas, Band 8 Charadriiformes (3. Teil). AULA-Verlag, Wiesbaden.

Grunsky-Schöneberg B (1998). Brutbiologie und Nahrungsökologie der Trottellumme (*Uria aalge* Pont.) auf Helgoland. Ökol. Vögel 20:217-274.

Hastie T & Tibshirani R (1990). Generalised Additive Models. Chapman and Hall, London.

Hays GC (2003). A review of the adaptive significance and ecosystem consequences of zooplankton diel vertical migration. In: Jones MB, Ingólfsson EO, Ólafsson E, Helgason GV, Gunnarsson K & Svavarsson J (Eds.): Migrations and dispersal of marine organisms. Hydrobiologia 503:163-170.

Hunt GLJ & Schneider DC (1987). Scale-dependent processes in the physical and biological environment of marine birds. In: Croxall JP (ed.) Seabirds - feeding ecology and role in marine ecosystems. Cambridge University Press, New York, pp. 7-41.

Kahlert J, Desholm M, Clausager I, Petersen IK (2000). Environmental impact assessment of an offshore wind farm at Rødsand: Technical report on birds, Report from NERI: 65 pp.

Kellermann A, Eskildsen K, Frank B (2006). The MINOS project: ecological assessments of possible impacts of offshore wind energy projects. In: von Nordheim H, Boedeker D & Krause JC (Eds.): Progress in marine conservation in Europe. Springer, Berlin Heidelberg.

Laake JL, Buckland ST, Anderson DR, Burnham KP (1993). DISTANCE user's guide v2.0. Colorado Cooperative Fish & Wildlife Research Unit, Colorado State University, Fort Collins, CO.

Manikowski S (1971). The influence of meterological factors on the behaviour of seabirds. Acta Zool. Cracov. 20:581-667.

Markones N (2007). Habitat selection of seabirds in a highly dynamic coastal sea: temporal variation and influence of hydrographic features. Ph.D. thesis, University of Kiel.

Mönkkönen M & Aspi J (1997). Sampling error in measuring temporal density variability in animal populations and communities. Ann. Zool. Fennici 34:47-57.

Nettleship DN & Gaston AJ (1978). Patterns of pelagic distribution of seabirds in western Lancaster Sound and Barrow Strait, Northwest Territories, in August and September 1976. Canadian Wildlife Service, Ottawa, Occasional paper 39.

Pennycuick CJ (1997). Actual and 'optimum' flight speeds: field data reassessed. J. Exp. Biol. 200:2355-2361.

Petersen W, Petschatnikov M, Wehde H, Schroeder F (2005). FerryBox: Real-time monitoring of water quality by ferryboats. Environ. res. eng. manage. 3:12-17.

Shealer DA (2002). Foraging behavior and food of seabirds. In: Schreiber EA & Burger J (eds.): Biology of marine birds. CRC Press, Boca Raton, pp. 137-177.

Skov H & Prins E (2001). Impact of estuarine fronts on the dispersal of piscivorous birds in the German Bight. Mar. Ecol. Prog. Ser. 214:279-287.

Speckman SG, Springer AM, Piatt JF, Thomas DL (2000). Temporal variability in abundance of Marbled Murrelets at sea in southeast Alaska. Waterbirds 23:364-377.

Tasker ML, Hope Jones P, Dixon T, Blake BF (1984). Counting seabirds at sea from ships: a review of methods employed and a suggestion for a standardized approach. Auk 101:567-577.

Thiel R, Sepúlveda A, Kafemann R, Nellen W (1995). Environmental factors as forces structuring the fish community of the Elbe estuary. J. Fish Biol. 46:47-69.

Valenzuela G, Alheit J, Coombs SH, Knust R (1991). Spawning patterns of Sprat and survival chances of Sprat larvae in relation to frontal systems in the German Bight. ICES Doc. CM 1991/L:45.

Wanless S, Harris MP, Morris JA (1988). The effect of radio transmitters on the behavior of Common Murres and Razorbills during chick rearing. Condor 90:816-823.

Webb A & Durinck J (1992). Counting birds from ship. In: Komdeur J, Bertelsen J, Cracknell G (Eds.): Manual for aeroplane and ship surveys of waterfowl and seabirds. IWRB Special Publication: 24-37.

Wood SN (2000). Modelling and smoothing parameter estimation with multiple quadratic penalties. J. R. Statist. Soc. B 62:413-428.

Wood SN (2006). Generalized Additive Models. An Introduction with R. Chapman and Hall, London.

Woodcock AH (1940). Observations of Herring Gull soaring. Auk 57:219-224.

Excursus 6: Great Cormorant

Katrin Wollny-Goerke, Kai Eskildsen, Nele Markones, Stefan Garthe

The **Great Cormorant** (*Phalacrocorax carbo*) is a big seabird of about 90cm body length. When standing, it has a characteristic silhouette with a long body and very short legs. It can often be observed resting with spread wings to dry them after diving bouts. Its feet are webbed and formed like small paddles. Its plumage is nearly black, in the breeding period with a white throat and a white patch on the thigh, the wings are dark brown. The bill is long and lightly hooked at the end, with a yellow spot on the lower mandible.

Cormorants feed opportunistically on benthopelagic fish species (see chapter 8). They are good divers who move forward by paddling with their webbed feet. They mostly stay underwater for 15-60 seconds while performing dives to the sea bottom. Diving depths usually amount up to few metres but depths of more than 30m have been recorded in marine habitats.

Sometimes, they aggregate in large feeding flocks (see chapter 8).

Great Cormorants breed in large colonies on the coasts of the Baltic Sea and in smaller colonies in the Wadden Sea of the German North Sea coast. This seabird species stays in coastal waters as it needs resting places to dry its wings after foraging trips.
Cormorants belong to the migratory species regularly occurring in the German North and Baltic Seas, whose breeding, moulting, resting, migration and wintering areas have to be protected according to the EU Birds Directive.

8 Spatio-temporal patterns of inshore and offshore foraging by Great Cormorants in the southwestern Baltic Sea

Stefan Garthe, Nils Guse, Nicole Sonntag

Zusammenfassung

In diesem Artikel wird die Verbreitung des Kormorans (*Phalacrocorax carbo*) in der südwestlichen Ostsee in räumlicher und zeitlicher Hinsicht untersucht, ebenso wie die Nahrung dieser Vogelart. Während der Brutzeit waren Kormorane eindeutig am häufigsten im Greifswalder Bodden und angrenzenden Gebieten östlich von Rügen sowie nördlich von Usedom anzutreffen. Ein weiterer Verbreitungsschwerpunkt, wenn auch bei deutlich geringerer Häufigkeit, wurde in der Kieler Bucht ermittelt. Die Lage der Brutkolonien konnte die Verbreitungsschwerpunkte im Greifswalder Bodden nebst östlich angrenzenden Bereichen sowie in der Kieler Bucht gut erklären. Sämtliche Konzentrationsbereiche auf See befanden sich innerhalb einer Entfernung von 25 km von der nächstgelegenen Kolonie. Bezogen auf die Verbreitung um Rügen zeigten Kormorane saisonale Verschiebungen im regionalen Rahmen, wobei u.a. eine deutliche Konzentration im Greifswalder Bodden im Frühjahr zum Zeitpunkt des Laichens der Heringe auftrat. Nahrungssuchende Kormorane wurden häufig in großen Schwärmen bis zu 2.000 Individuen beobachtet. Derartige Schwärme wurden bis zu einer Distanz von 21 km von der nächstgelegenen Küstenlinie gesichtet und waren häufig von Silbermöwen, Mantelmöwen und anderen Arten begleitet. Mageninhaltsanalysen von neun auf Usedom eingesammelten Kormoranen ergaben, dass eine ganze Reihe von Fischarten und -familien erbeutet wurden, wobei erhebliche individuelle Variabilität in der Biomasse auftrat. Kaulbarsch und Rotauge, zwei kommerziell unbedeutende Arten, waren die bei weitem wichtigste Beute und trugen 69 % zur erbeuteten Biomasse bei. Dies belegt wie schon bei älteren Studien, dass sich Kormorane opportunistisch ernähren und große regionale Unterschiede in ihrer Nahrungswahl aufweisen. Insgesamt wird aus dieser Arbeit deutlich, dass der Kormoran eine häufige Art im küstennahen Bereich der südwestlichen Ostsee ist und große Ansammlungen in den Bodden und küstennahen Flachwasserbereichen aufweist. Dies zeigt die große Bedeutung dieser Gebiete und macht sie somit z. B. als Windpark-Standort ungeeignet. Großräumige Untersuchungen wie im MINOS-Projekt sind notwendig, um die Bedeutung der verschiedenen Seegebiete für die zahlreichen Seevogelarten zu bewerten.

Abstract

In this article we analyse the spatial and temporal distribution as well as the diet of Great Cormorants (*Phalacrocorax carbo*) in the southwestern Baltic Sea. During the breeding period, foraging Cormorants were clearly most abundant in the Greifswald Lagoon and nearby areas east of Rügen and north of Usedom. A further centre of distribution, although with much lower abundances, was found in the Kiel Bight. Colony distribution explains well the overall concentration of birds in/east of the Greifswald Lagoon and the Kiel Bight. All concentration areas at sea are located within 25km of the nearest colony. Focusing on the area around Rügen, Cormorants show seasonal shifts at a regional level, with an apparent concentration in the Greifswald Lagoon in spring when herring is spawning in that area. Large flocks of feeding Cormorants were frequently observed, with flock sizes of up to 2,000 individuals. Such flocks were found up to 21km off the nearest coastline. Flocks of Cormorants were often accompanied by Herring Gulls, Great Black-backed Gulls and several other species. From stomach analysis of nine Cormorants from the island of Usedom it could be derived that a range of fish species and families was preyed upon, with extensive individual variation in prey biomass. Ruffe and Roach, commercially unimportant species, were by far the most important prey, contributing almost 69% of the total prey biomass. Cormorants are opportunistic feeders that show large regional differences in their diet. It was concluded that the Great Cormorant is a numerous bird in inshore areas of the southwestern Baltic Sea, with high aggregations in the lagoons and shallow areas relatively close to the coast. This highlights the importance of such areas and makes them unsuitable as sites for offshore wind farms. Large-scale surveys such as those conducted within the MINOS project are needed to evaluate the importance of different sea areas for the various seabird species.

Introduction

The Baltic Sea is a sea area of particular conditions, most evident by the strong horizontal and vertical salinity gradients (Rheinheimer 1995). This brackish sea hosts a special avifauna with large numbers of waterbirds while highly pelagic seabird species are virtually absent (e.g. Durinck et al. 1994, Garthe et al. 2003). In contrast to other more marine seas such as the North Sea and the Mediterranean Sea, freshwater species regularly occur in the offshore areas of the Baltic Sea. Besides e.g. mergansers, grebes and sea ducks, the Great Cormorant (*Phalacrocorax carbo*, named Cormorant subsequently) is one of these species that occur in both freshwater and marine habitats. As has been shown by previous publications derived from the MINOS project and others (Garthe et al. 2003, 2004, Sonntag et al. 2006), the coastal waters of the Baltic Sea have major relevance for this species that needs to be investigated in more detail. The Cormorant is also a concern of fishermen and nature conservationists as the debate on the possibly adverse influence of food consumption by Cormorants on commercial fish stocks continues (see e.g. recent overview by Herzig & Böhnke 2007). In terms of planning offshore wind farms, Cormorants have largely been ignored. However, they were ranked highly vulnerable by wind turbines (Garthe & Hüppop 2004) and hence their distribution at sea is an important issue. The information in this article on the functionality of the Baltic Sea for the Cormorant and on its diet should contribute to the evaluation of the role of this native breeding species in the marine ecosystem.

Material and Methods

Distribution of Cormorants at sea was analysed using ship-based seabirds-at-sea transect counts following the methodology of Tasker et al. (1984) and Garthe et al. (2002). Data originated from the FTZ ship database version 5.09, as of October 2007. Observations in sea states higher than 5 (see description by Garthe et al. 2002) were excluded as conditions were considered too unfavourable for properly counting Cormorants. The data analysed for this study are comprised of more than 49,800km of ship transects from 2000 to 2007. In distribution maps, abundance is given as the number of birds counted in/on the water per distance traveled. This also includes birds outside the sampling transect but excludes birds that were only observed flying. Distribution maps are based on ordinary kriging interpolation procedures as described by Garthe (2003).

The diet of Cormorants was investigated in nine birds from the island of Usedom and surrounding waters, most of which were by-caught in set nets off the Usedom coast (around the 10 m depth contour). Most birds were immature and originated from different seasons of the years 2001 to 2003. Six of the birds were in good or moderate condition, three in poor condition. The analysis of stomach contents followed the methodology described by Leopold et al. (1998), Ouwehand et al. (2004) and Guse (2005). The length of prey fish was either directly measured or reconstructed from measurements of hard parts (otoliths, chew pads). Original fresh biomass was calculated mostly based on regressions from Leopold et al. (2001), accounting for the progressive wear of dietary remains due to the digestion process.

Results

Distribution and behaviour

During the breeding period, Cormorants on/in the water were clearly most abundant in the Greifswald Lagoon and nearby areas east of Rügen and north of Usedom (Fig. 1).

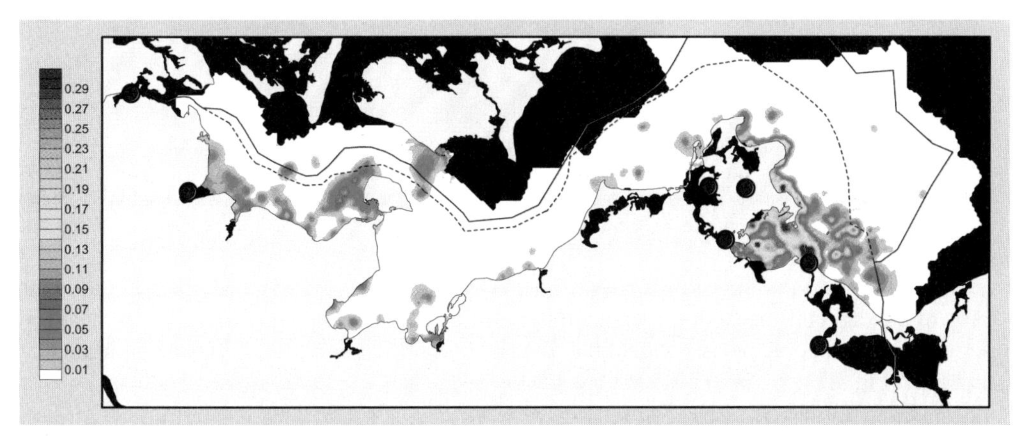

Figure 1: Distribution of Cormorants in the southwestern Baltic Sea during the breeding period (April to July), 2000-2007. The colour scale gives the abundance as birds/km at a logarithmic scale. Black colour indicates areas not studied. Only birds in/on the water are shown. The circles indicate the location and size of coastal breeding colonies with more than 100 pairs in 2005 (as taken from Heinicke 2005 and Koop & Kieckbusch 2005).

A further centre of distribution, although with much lower abundances, was found in the Kiel Bight. Colony distribution explained well the overall concentration of birds in/east of the Greifswald Lagoon and the Kiel Bight. The much higher abundances in the eastern study area fit well with the much higher number of breeding birds (Fig. 1). Although the status of the birds at sea was generally not determined, it seems likely that most Cormorants were related to the breeding colonies. In accordance with this assumption, all concentration areas at sea were located within 25km of the nearest colony (Fig. 1).

Focusing on the core region around Rügen Island, it can be seen that concentrations of Cormorants also occurred in this area during other periods of the year. After the breeding season, during autumn migration, numbers in the Greifswald Lagoon strongly diminished, while the sea waters off Usedom remained important (Fig. 2A).

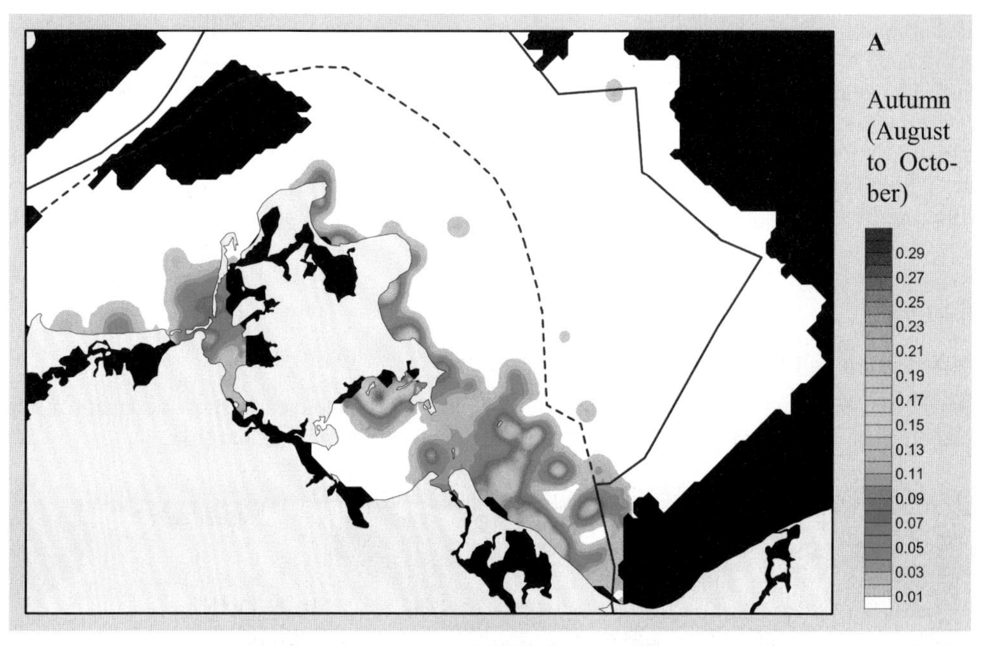

Figure 2 A: Distribution of Cormorants around Rügen in the southwestern Baltic Sea outside the breeding period, 2000-2007. See figure 1 for more details. A - Autumn

In winter, abundances at sea decreased in general, while they increased slightly in the inner parts of the lagoon (Fig. 2B). Coverage of the Baltic Sea in spring was lower than in the other periods, but it is nevertheless evident that the Greifswald Lagoon was of major importance at that time (Fig. 2C).

The distribution patterns shown in Fig. 1 and 2 reflect the distribution of foraging birds: 95% of all birds with behaviour recordings (n = 1921 individuals) exhibited foraging behaviour while only 5% were either bathing, preening or resting. When not foraging, birds were mostly associated with set nets and other fishing gear, poles and other structures in the open water. No associations were recorded with fishing boats except for one anecdotal observation of 16 individuals following a set net boat near Fehmarn on 14 January 2001 (O. Hüppop pers. comm).

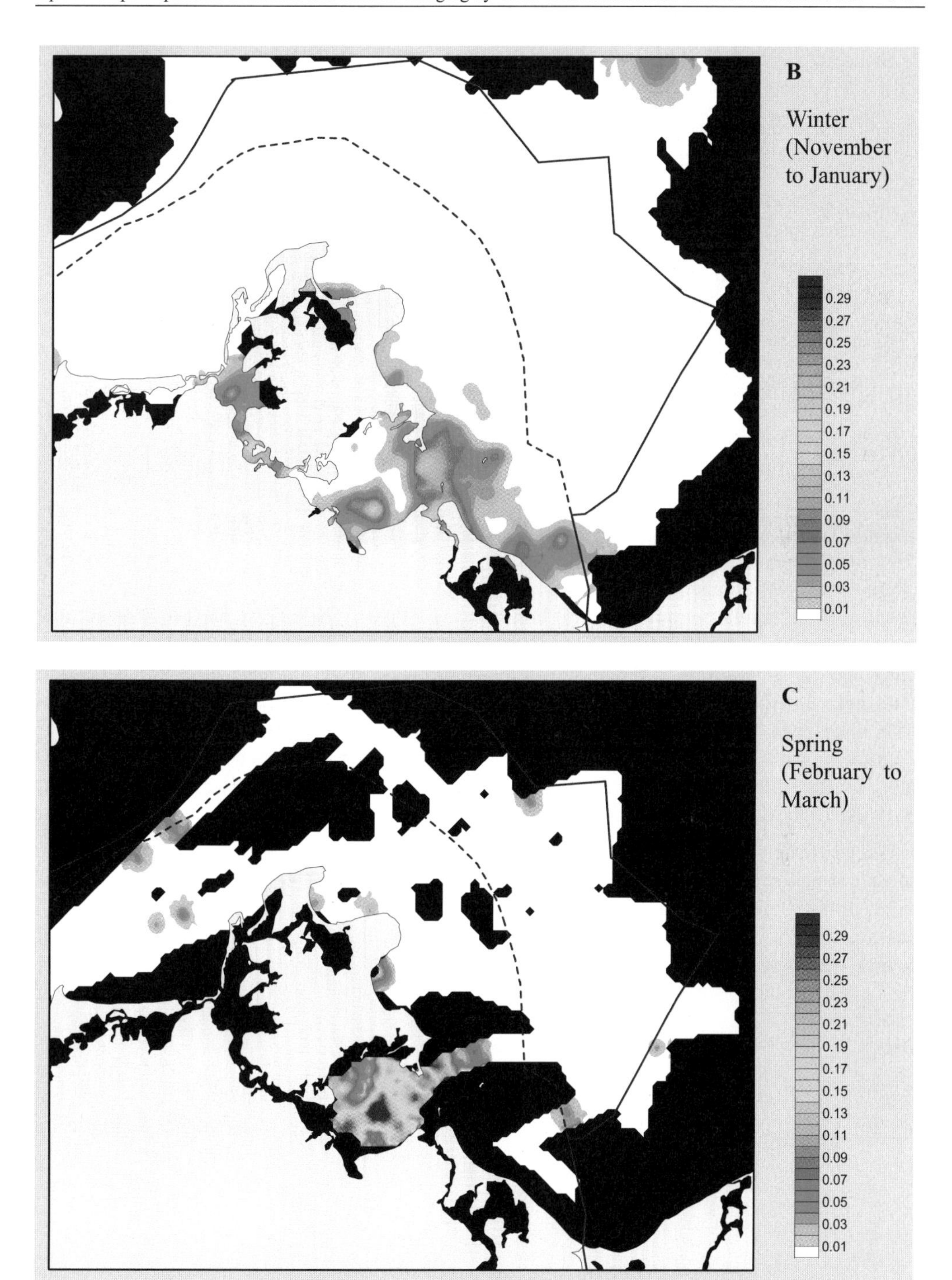

Figure 2 B & C: Distribution of Cormorants around Rügen in the southwestern Baltic Sea outside the breeding period, 2000-2007. See figure 1 for more details. B – Winter, C – Spring.

Figure 3:

Feeding flocks of Cormorants fishing in the southwestern Baltic Sea

Cormorants often foraged in groups at sea (Fig. 3).
Feeding flocks could comprise several hundred individuals. The largest flocks observed during the ship-based surveys were seen 3km from the coast east of Rügen Island (Göhren, 2,000 ind., 25 July 2003), at the mouth of the river Peene in the southern Greifswald Lagoon (700 ind., 25 April 2002) and in the Wismar Bay (600 ind., 20 October 2004). The month with most large flocks was April, and the area with most large flocks was the Greifswald Lagoon region. Large flocks were also seen in the open sea, e.g. 10km north of Usedom (230 ind., 29 June 2004) and at the southwestern edge of the Oderbank, 21km off the nearest coastline (370 ind., 18 May 2005).

Especially when Cormorants formed flocks, other bird species often associated with them. Most of such multi-species feeding associations existed with Herring Gulls (*Larus argentatus*, up to 50 per flock) and Great Black-backed Gulls (*Larus marinus*, up to 20 per flock), but also with Common Gulls (*Larus canus*), Common Eiders (*Somateria mollissima*) and several other species.

Diet

A range of fish species and families was found in the nine Cormorants investigated (Table 1).

Table 1: Diet composition of nine Cormorants collected at Usedom based on stomach samples. See text for details.

Group	Species/Category	Occurrence (n = birds)	Number of prey items	Proportion of all prey items (in %)	Proportion of total prey biomass (%)
Fish	Ruffe (*Gymnocephalus cernuus*)	2	24	25.0	42.5
Fish	Roach (*Rutilus rutilus*)	2	13	13.5	26.4
Fish	Perch (*Perca fluviatilis*)	1	1	1.0	5.7
Fish	Fish indet.	3	4	4.2	5.6
Fish	Atlantic Herring (*Clupea harengus*)	1	1	1.0	5.0
Fish	Perch indet. (Percidae)	3	3	3.1	4.7
Crustacea	Shrimp indet. (Palaemonidae)	1	30	31.3	2.8
Fish	Sprat (*Sprattus sprattus*)	1	2	2.1	1.8
Fish	Rudd (*Scardinius erythrophthalmus*)	1	1	1.0	1.4
Fish	Sandeel indet. (Ammodytidae)	2	4	4.2	1.4
Fish	Zander (*Sander lucioperca*)	1	1	1.0	1.2
Fish	Eel (*Anguilla anguilla*)	1	1	1.0	0.8
Fish	Three-spined Stickleback (*Gasterosteus aculeatus*)	1	3	3.1	0.4
Fish	Smelt (*Osmerus eperlanus*)	2	6	6.3	0.1
Fish	Goby indet. (*Pomatoschistus spp.*)	1	2	2.1	0.1

Extensive individual variation in prey biomass was found. Ruffe (*Gymnocephalus cernuus*) and Roach (*Rutilus rutilus*) were by far the most important prey, contributing almost 69% of the total prey biomass. This result was corroborated by the fact that the only birds exhibiting full stomachs had fed extensively on these two non-commercial fish species (Fig. 4).

In two birds, jaws of polychaete worms could be found. However, as these samples also contained numerous prey fish, it was unclear whether they had been targeted as primary prey, as described by Leopold & van Damme (2002), or were prey of the fish eaten by the Cormorants. In one single bird remains of 30 shrimps (*Palaemonidae*) could be found and thus were most probably consumed as primary prey.

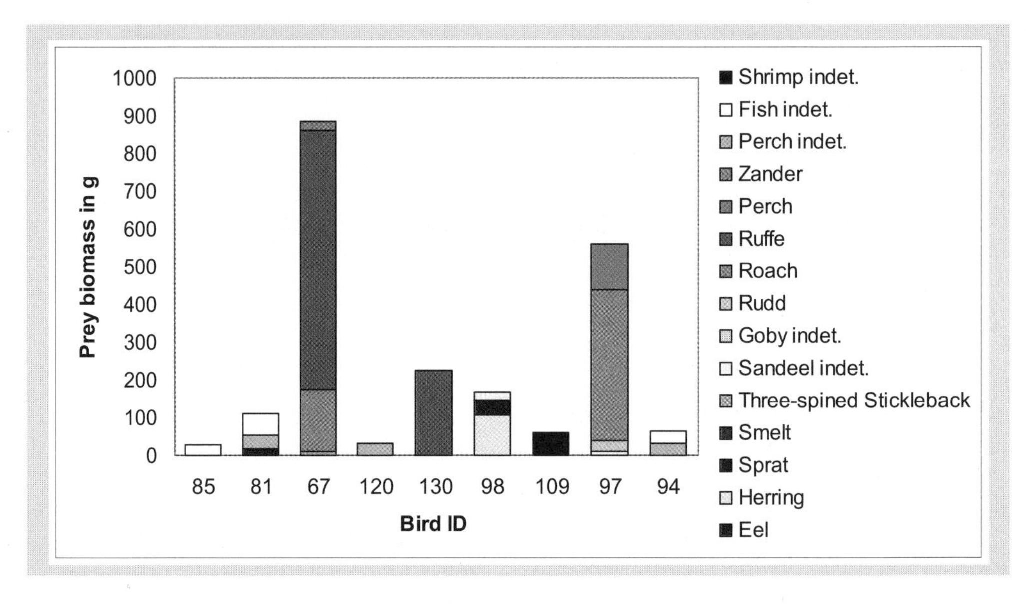

Figure 4: Calculated prey biomass for nine Cormorants, based on stomach analysis. See text for details.

Discussion

Cormorants primarily use the inshore areas of the southwestern Baltic Sea for foraging but some areas further offshore are also frequently used. This fits well the pattern known from many benthivorous or bentho-pelagic-feeding waterbird species that shallow grounds are of major importance, especially for sea ducks, but also for grebes and others (e.g. Bräger & Nehls 1987, Sonntag et al. 2006). In contrast to e.g. Common Scoters (*Melanitta nigra*), Velvet Scoters (*Melanitta fusca*), Long-tailed Ducks (*Clangula hyemalis*), Slavonian Grebes (*Podiceps auritus*) and Red-necked Grebes (*Podiceps grisegena*), Cormorants do not use the Oderbank in the Pomeranian Bight for foraging, although water depths would be suitable. The frequent use of shallow grounds in the Kiel Bight but not in the central and outer Pomeranian Bight highlights that this species tends to stay close to the coast, similar to e.g. Great Crested Grebes (*Podiceps cristatus*) or Red-breasted Mergansers (*Mergus serrator*) (Sonntag et al. 2006).

Cormorants need high prey densities when foraging (Grémillet & Wilson 1999). The high abundances in areas with high fish quantities confirm this pattern. The concentration of Cormorants in the Greifswald Lagoon in spring and during the beginning of the breeding season matches the main spawning season of Herring (*Clupea harengus*) in this area, which usually lasts from March to May (Klinkhardt 1996). Many sightings of mass feeding events of Cormorants consisting of flocks of several hundred individuals could be explained by social fishing that enables the birds to exploit abundant schooling fish species in turbid, euthrophic environments (van Eerden & Voslamber 1995).

The diet composition of the nine Cormorants analysed in this study seem to reflect the coastal fish fauna of Usedom Island, containing freshwater, brackish and a few marine species. Especially Ruffe is known to prefer brackish habitats. Both main prey fish species, Ruffe and Roach, prefer eutrophic conditions. Eutrophication processes over the last decades have been shown to increase production of water bodies and thus the biomass of small-sized, early-mature fish species such as Roach and Ruffe (de Nie 1995), leading to an improved food supply for Cormorants which seems to be one of the main reasons for the increase of the Cormorant population in the recent past (van Eerden & Gregersen 1995). The extensive individual variation in prey biomass found was very likely generated or at least enhanced by varying factors such as timing of capture/death (health status), age, sex, condition, year and season (see Guse 2005). Cormorants are diet opportunists, reflecting the local fish fauna and its spatial, seasonal and interannual differences in terms of abundance (Knief 1994). By analysing 237 pellets, Preuß (2002) showed that sticklebacks, sandeels and Roach were most important prey of Cormorants breeding in a colony at the Baltic Sea coast close to our study area between 1997 and 1999. Studies by the Vogelwarte Hiddensee between 1959-1968 showed that Cormorant stomachs contained 30% of both Herring and Perch and 20% eel (*Anguilla anguilla*; Strunk & Strunk 2005). Even though our small sample size must be treated with caution, it is remarkable that only a single Eel, which contributed less than 1% of the total prey biomass, could be found in our study. This difference to the data of Strunk & Strunk (2005) may be due to the overall decline of the European Eel stock in recent decades, making Eel a less abundant prey. The recruitment of this species has dropped to 1% of former levels and the stock is considered to be dangerously close to collapse (www.ices.dk/marineworld/eel.asp). Ubl (2004) analysed fish found in stomachs of 83 Cormorants shot in the Greifswald Lagoon. He found Herring (55%), Three-spined stickleback (*Gasterosteus aculeatus*; 11%) and Perch (10%) to be the most important prey by biomass. However, there was strong variation over time. The maximum consumed biomass of 878 g reconstructed for one bird of our study seems to clearly exceed the daily need described for Cormorants (Kube 2004). As the degradation of dietary remains is a function of their size and as chew pads and large otoliths as from Ruffe may remain in the stomachs for more than one day, the biomass found most likely refers to a food consumption of more than one day.

An interesting side aspect of the feeding ecology of the Cormorant in the Baltic Sea is that this species does not exploit fish from fishing trawlers, neither from the net when it is trawled nor from the discards discharged when sorting the catch, as is the case in the southern North Sea (Camphuysen 1999). The widespread occurrence of free-ranging fish as well as set nets and other static fishing gear might be more attractive for Cormorants. Also, trawl fisheries in the Baltic Sea usually occur further offshore.

In conclusion, it could be shown that the Great Cormorant is a numerous bird in inshore areas of the southwestern Baltic Sea, with high aggregations in the lagoons and shallow areas relatively close to the coast. This highlights the importance of such areas and makes them unsuitable as sites for offshore wind farms. Large-scale surveys such as those conducted within the MINOS project are needed to evaluate the importance of different sea areas for the various seabird species. With such large-scale surveys, ideally in combination with detailed behavioural observations of seabirds at sea and with diet studies, the ecology of species can be comprehensively investigated.

Acknowledgements

Most ship-based surveys and data analyses were financially support by the German Ministry for Environment, Nature Conservation and Nuclear Safety (BMU) and the German Federal Agency for Nature Conservation (BfN) as well as the Ornithologische Arbeitsgemeinschaft für Schleswig-Holstein & Hamburg e.V. Many observers assisted in bird counts over the years. B. Schirmeister provided the Cormorants for diet analyses. P. Schwemmer gave valuable comments on an earlier version of the manuscript.

References

Bräger S & Nehls G (1987). Die Bedeutung der schleswig-holsteinischen Ostsee-Flachgründe für überwinternde Meeresenten. Corax 12:234-254.

de Nie H (1995). Changes in the inland fish populations in Europe in relation to the increase of the Cormorant *Phalacrocorax carbo sinensis*. Ardea 83:115-122.

Durinck J, Skov H, Jensen FP, Pihl S (1994). Important marine areas for wintering birds in the Baltic Sea. Ornis Consult report, Kopenhagen.

Garthe S (2003). Verteilungsmuster und Bestände von Seevögeln in der Ausschließlichen Wirtschaftszone (AWZ) der deutschen Nord- und Ostsee und Fachvorschläge für EU-Vogelschutzgebiete. Ber. Vogelschutz 40:15-56.

Garthe S, Dierschke V, Weichler T, Schwemmer P (2004). Rastvogelvorkommen und Offshore-Windkraftnutzung: Analyse des Konfliktpotenzials für die deutsche Nord- und Ostsee. Final report for the sub-project 5 within the project "Marine Warmblüter in Nord- und Ostsee: Grundlagen zur Bewertung von Windkraftanlagen im Offshorebereich (MINOS)".

Garthe S & Hüppop O (2004). Scaling possible adverse effects of marine wind farms on seabirds: developing and applying a vulnerability index. J. Appl. Ecol. 41:724-734.

Garthe S, Hüppop O, Weichler T (2002). Anleitung zur Erfassung von Seevögeln auf See von Schiffen. Seevögel 23:47-55.

Garthe S, Ullrich N, Weichler T, Dierschke V, Kubetzki U, Kotzerka J, Krüger T, Sonntag N, Helbig AJ (2003). See- und Wasservögel der deutschen Ostsee - Verbreitung, Gefährdung und Schutz. Bundesamt für Naturschutz, Bonn.

Grémillet D & Wilson RP (1999). A life in the fast lane: energetics and foraging strategies of the Great Cormorant. Behav. Ecol. 10:516-524.

Guse N (2005). Diet of a piscivorous top predator in the Baltic Sea – the Red-throated Diver (*Gavia stellata*) in the Pomeranian Bight. Diploma Thesis, University of Kiel.

Heinicke T (2005). Zur Situation des Kormorans (*Phalacrocorax carbo sinensis*) in Mecklenburg-Vorpommern. Ber. Vogelschutz 42:97-122.

Herzig F & Böhnke A (eds.) (2007). Fachtagung Kormorane 2006. BfN-Skripten 204:1-243.

Klinkhardt M (1996). Der Hering. Spektrum Akademischer Verlag, Heidelberg, Berlin, Oxford.

Knief W (1994). Zum sogenannten Kormoran-„Problem“. Eine Stellungnahme der deutschen Vogelschutzwarten zum Kormoran – Bestand, Verbreitung, Nahrungsökologie, Managementmaßnahmen. Natur Landsch. 69:251-258

Koop B & Kieckbusch J (2005). Ornithologische Begleituntersuchungen zum Kormoran. Bericht für 2005. Unpubl. report for the Ministery for Environment, Nature Conservation and Agriculture in Schleswig-Holstein.

Kube J (2004). 50 Jahre Niederhof – die Geschichte einer deutschen Kormorankolonie. Falke 51:256-262.

Leopold MF & van Damme CJG. (2002). Great Cormorants *Phalacrocorax carbo* and polychaetes: can worms sometimes be a major prey of a piscivorous seabird? Mar. Ornithol. 31:83-87

Leopold MF, van Damme CJG, Philippart CJM, Winter CJN (2001). Otoliths of North Sea fish: interactive guide of identification of fish from the SE North Sea, Wadden Sea and adjacent fresh waters by means of otoliths and other hard parts. CD-Rom, ETI, Amsterdam.

Leopold MF, van Damme CJG, van der Veer HW (1998). Diet of Cormorants and the impact of Cormorant predation on juvenile flatfish in the Dutch Wadden Sea. J. Sea Res. 40:93-107.

Ouwehand J, Leopold MF, Camphuysen CJ (2004). A comparative study of the diet of Guillemots *Uria aalge* and Razorbills *Alca torda* killed during the Tricolor oil incident in the south-eastern North Sea in January 2003. Atlantic Seabirds 6:147-164.

Preuß D (2002). Nahrungsökologische Untersuchungen zu Einfluss des Kormorans *Phalacrocorax carbo sinensis* auf die Fischerei im Küstenbereich Vorpommerns. Naturschutzbeob. Mecklenburg-Vorpommern 45:57-67

Ubl C (2004). Untersuchungen zum Nahrungsspektrum des Kormorans im Bereich des Greifswalder Boddens. Fischerei Fischereimarkt Mecklenburg-Vorpommern 4:32-38.

Rheinheimer G (1995). Meereskunde der Ostsee. 2. Aufl., Springer, Berlin.

Sonntag N, Mendel B, Garthe S (2006). Die Verbreitung von See- und Wasservögeln in der deutschen Ostsee im Jahresverlauf. Vogelwarte 44:81-112.

Strunk G & Strunk P (2005). Die Entwicklung des Kormoranbestandes *Phalacrocorax carbo sinensis* am Strelasund und in der vorpommerschen Boddenregion. Meer und Museum 18:150-156.

Tasker ML, Hope Jones P, Dixon T, Blake BF (1984). Counting seabirds at sea from ships: a review of methods employed and a suggestion for a standardized approach. Auk 101:567-577.

van Eerden M & Gregersen J (1995). Long-term changes in the northwest European population of Cormorants. Ardea 83:61-79.

van Eerden MR & Voslamber B (1995). Mass fishing by Cormorants *Phalacrocorax carbo sinensis* at Lake IJsselmeer, The Netherlands, a recent and successful adaptation to a turbid environment. Ardea 83:199-212.

Glossary

Acoustic tag
Electronic data-logger for tagging animals to record the animal's (echolocation) sound

Agreement on the Conservation of Seals in the Wadden Sea
The Agreement on the Conservation of Seals in the Wadden Sea came into effect in Germany in 1990/1991 as a regional agreement under the → Bonn Convention, CMS. This is the most important Common seal protection agreement for Denmark, Germany and the Netherlands. The associated Seal Management Plan implements the agreement for the entire Wadden Sea. It also covers Grey seals.

Air-borne sound
Sound propagating in air

Amplitude
Maximum absolute value of a wave measured along its vertical axis

ASCOBANS
Agreement on the Conservation of Small Cetaceans of the Baltic and North Seas.
This Agreement on the Conservation of Small Cetaceans of the Baltic and North Seas was signed in 1991 and has been legally binding in Germany since 1994. ASCOBANS is an agreement supplementary to the Convention on the Conservation of Migratory Species of Wild Animals (→ Bonn Convention, CMS). Under ASCOBANS, the Federal Republic has an obligation to undertake research and protection measures to secure the preservation of this whale species.

Aversive
Unwilling

Biosonar
Echolocation (see below)

Bonn Convention
The 1979 Convention on the Conservation of Migratory Species of Wild Animals (CMS, the Bonn Convention) protects migratory wild animal species endangered worldwide and regionally.

Broadband sound
Sound consisting of a wide band of frequencies

Click impulse
Sharp transient sound pressure wave

Data logger
Small microchip unit with a memory capacity up to some dozens MB which records information on different sensors in defined intervals (e.g. 3-20 seconds), mostly put in a device of synthetic material.

Dead-reckoning system
Satellite supported data logger which provides continuous records of all important activities of the seals on land and in the water for periods up to 2-3 months, corporated in a pressure-resistant positively buoyant body which releases itself after a prescribed time. It is fitted on the seal in a sort of "backpack" base, made of neoprene, which comes off during the annual moulting. After retrieval, the stored data can be downloaded and analysed by special software.

Detection threshold
Lowest sound pressure level that can be detected

Discards
Fish and invertebrates which are caught at sea and which are subsequently discharged again because they cannot be commercially exploited

Echolocation
Active sensory system in certain animals: The emission of sound by an animal and perception of its echo to determine a bearing, range or the characteristics of an echoing object

Foraging trip
A period of time spent at sea searching for food.

Frequency
The number of oscillations or cycles per unit of time

Front (here: hydrographic front)
Steep gradient of hydrographical characteristics such as temperature or salinity over a short horizontal distance; caused by different flow velocities of neighbouring water bodies; can often be recognised due to an abrupt change in water colour and aggregations of foam, biomass or litter along the front; therefore enhanced primary production and passive aggregation of prey organisms.

g(0)
Probability of detection on the transect line, usually assumed to be 1. In the case of marine mammals that spend substantial periods underwater and thus avoid detection, this parameter must be estimated based on other type of information (Buckland et al. 2001).

Habitat
A locality which is separated from others or preferable used due a certain (internal or external) parameter

Haul-out location
Areas on sand banks or on coasts that are used by seals for resting, pupping, or moulting

Hydrography
Here: refers to physical attributes of water bodies including temperature-salinity structure, gradients, → hydrographic fronts, mixing, waves, tides and currents.

Hydrophone
Submersible sound sensor for picking up underwater sound

Larynx
Upper part of the trachea in most vertebrate animals, containing the vocal cords

Melon
Oval shaped oily, fatty lump of tissue found at the centre of the forehead of most dolphins and toothed whales.

Physical
Concerned with the body of an organism (or organs) and its needs

Physiological
Relating to the normal vital functions and activities of life or of living matter

Phonic lips
Structure located in the passages of the soft-tissue nasal complex superior to the skull in toothed whales

Satellite tag
Electronic data-logger for tagging animals to track their movements via satellite

SCANS
Small Cetacean Abundance in the North Sea and Adjacent Waters

Sensory system
Part of the nervous system responsible for processing sensory information

Sound pressure level
20 times the logarithm to the base 10 of the ratio of a sound pressure to a named reference pressure (unit: → decibel, dB)

Telemetry
Remote collection of data using radio wave transmission, acoustic signals, satellites, or data loggers

Transect
Survey route or subsample of the study area which is surveyed to gain a representative dataset used for extrapolating information to the level of the whole study area

Abbreviations:

AEPs
Auditory evoked potentials; Electrical potentials generated within neuronal nuclei at different positions in the auditory system upon the perception of sound stimuli.

dB
Decibel; A dimensionless ratio term that can be applied to any two values → Sound pressure level

EEZ
Exclusive Economic Zone: area between the territorial sea (12nm-zone) and the high seas, normally 12nm to 200nm.

esw
Effective half-strip width of the area searched effectively on each side of the line transect (Buckland et al. 2001)

FTZ
Forschungs- und Technologiezentrum Westküste Büsum = Research and Technology Centre Westcoast, Büsum

GOM
German Oceanographic Museum, Stralsund

Hz
Abbreviation for Hertz. Unit for the frequency of a periodic procedure, 1 oscillation per second.

IWC
International Whaling Commission

kHz
Abbreviation for Kilohertz, 1,000 Hertz

MTTS
Masked Temporary Threshold Shift → TTS

PDV
The Phocine Distemper Virus affects the seal´s immune system and makes them highly susceptible to secondary infections such as pneumonia. If a seal is infected it will very likely die.

PPD
Porpoise Positive Day: a monitored day with porpoise registration

PTS
Permanent Threshold Shift (→ TTS); permanently impaired perception of sound due to decreased hearing sensitivity

SEL
Acoustic energy over time, measured in dB re → μPa

T-POD
Timing Porpoise Detector: a submersible data-logger that detects echolocation sound of toothed whales

TTS
Temporary Threshold Shift. A degradation of the hearing sensitivity, mostly caused by a too strong acoustic irradiation, which is however reversible. If 30 days after the acoustic irradiation a deviation from the normal condition is still present, it is defined as → PTS.

μPa
Unit for the acoustic pressure / excess pressure

Publications in the frame of MINOS Research Network

General

Federal Maritime and Hydrographic Agency (BSH)(2007). Standard Untersuchung der Auswirkungen von Offshore-Windenergieanlagen auf die Meeresumwelt (StUK 3), 58 pp.

Frank B (2006). Research on Marine Mammals. Summary and discussion of research results. In: Köller J, Köppel J, Peters W (Eds.). Offshore wind energy – Research on Environmental impacts. Springer, Berlin, Heidelberg, pp. 77-86

Kellermann A, Eskildsen K, Frank B (2006). The MINOS-project: ecological assessments of possible impacts of offshore wind energy projects. In: von Nordheim H, Boedecker D, Krause JC (Eds.): Progress in Marine Conservation in Europe - Natura 2000 sites in German Offshore Waters. Springer, Berlin, Heidelberg, pp. 239-248

Kellermann A (2004). Offshore-Windparks und die Meeresumwelt.- In: Altner G, Leitschuh-Fecht H, Michelsen G, Simonis UE, von Weizsäcker EU (Eds.). Jahrbuch Ökologie 2005. C.H. Beck, München, pp. 74-82

Kellermann A, Eskildsen K, Frank B (Eds.) (2004). MINOS – Marine Warmblüter in Nord- und Ostsee: Grundlagen zur Bewertung von Windkraftanlagen im Offshore-Bereich. Final report Vol. I und II, research network (FKZ 032750) financed by Federal Ministry of Environment, Nature protection and nuclear safety, unpubl., 460 pp.

Eskildsen K, Adelung D, Benke H, Dehnhardt G, Garthe S, Kellermann A, Lucke K, Scheidat M, Siebert U, Verfuß U, Wilson R (2002). MINOS – Marine Warmblüter in Nord- und Ostsee: Grundlagen zur Bewertung von Windkraftanlagen im Offshore-Bereich. In: Kutscher, J (Ed.): Ökologische Begleitforschung zur Offshore-Windenergienutzung Tagungsband, Bremerhaven 28.-29.05.2002. - PTJ, Jülich, pp. 17-29

www.minos-info.de - german

www.minos-info.org - english

Harbour porpoise

Lucke K, Lepper PA, Hoeve B, Everaarts E, van Elk N, Siebert U (2007). Perception of Low-Frequency Acoustic Signals by a Harbour Porpoise (*Phocoena phocoena*) in the Presence of Simulated Offshore Wind Turbine Noise. Aquatic Mammals, 33(1):55-68.

Verfuß UK, Honnef CG, Meding A, Dähne M, Mundry R, Benke H (2007). Geographical and seasonal variation of harbour porpoise (*Phocoena phocoena*) presence in the Baltic Sea revealed by passive acoustic monitoring. Journal of the Marine Biological Association of the United Kingdom 87:165-176

Madsen PT, Wahlberg M, Tougaard J, Lucke K, Tyack P (2006). Wind turbine underwater noise and marine mammals: Implications of current knowledge and data needs - Review. Marine Ecology Progress Series, 309:279-295.

Scheidat M, Gilles A, Siebert U (2006). Evaluating the distribution and density of harbour porpoises (*Phocoena phocoena*) in selected areas in the German waters. In: v. Nordheim H, Boedeker D, Krause J (Eds.): Progress in marine conservation in Europe - NATURA 2000 sites in German offshore waters, Springer, Hamburg, pp. 189-208

Siebert U, Benke H, Dehnhardt G, Gilles A, Hanke W, Honnef CG, Lucke K, Ludwig S, Scheidat M, Verfuß UK (2006). Harbour porpoises (*Phocoena phocoena*): investigations of density, distribution patterns, habitat use and acoustics in the German North and Baltic Seas. In: Köller J, Köppel J, Peters W (Eds.). Offshore wind energy. Research on environmental impacts. Springer Verlag, Berlin, pp. 37-64.

Verfuß UK, Honnef CG, Benke H (2006): Seasonal and geographical variation of harbour porpoise (*Phocoena phocoena*) habitat use in the German Baltic Sea monitored by passive acoustic methods (PODs). In: v. Nordheim H, Boedeker D, Krause J (Eds.): Progress in marine conservation in Europe - NATURA 2000 sites in German offshore waters, Springer, Hamburg, pp. 209-224

Meding A (2005). Untersuchungen zur Habitatnutzung von Schweinswalen (*Phocoena phocoena*) in ausgewählten Gebieten der Ostsee mit Hilfe akustischer Methoden. Diplomathesis. Ernst-Moritz-Arndt-Universität Greifswald.

Verfuß UK, Miller LA, Schnitzler H-U (2005). Spatial orientation in echolocating harbour porpoises (*Phocoena phocoena*). Journal of Experimental Biology 208:3385-3394.

Herr H (2004). Sichtungshäufigkeit und Verbreitungsmuster von Meeressäugern in deutschen Gewässern - Relation zu Schifffahrts- und Fischereiaktivität. Examensarbeit, University of Hamburg, 76 pp.

Scheidat M, Kock K-H, Siebert U (2004). Summer distribution of harbour porpoise (*Phocoena phocoena*) in the German North and Baltic Sea. Journal of Cetacean Research and Management 6(3):251-257

Gilles A (2003). Verbreitungsmuster von Schweinswalen (*Phocoena phocoena*) in deutschen Gewässern. Diplomathesis, University of Kiel, 145 pp.

Kotzian S (2003). T-PODs – eine geeignete Methode das Verhalten von Schweinswalen (*Phocoena phocoena*) in ihrem natürlichen Lebensraum zu untersuchen? Diplomathesis, Universität Rostock.

In addition to these publications in journals and monographies, MINOS research network contributes with posters, expert papers and talks to the following prominent international conferences:

2007: 21st European Cetacean Society Conference in San Sebastian, Spain

Conference “Year of the Dolphin in Europe - Conservation of small cetaceans and marine protected areas” in Stralsund

2006: 13th ASCOBANS meeting

ICES Annual Science Conference in Maastricht, Netherlands

20th European Cetacean Society Conference in Gydnia, Poland

Static Acoustic Monitoring (SAM) Workshop as a tool for Environmental Impact Studies with emphasis on Offshore Wind Farm Constructions in Stralsund

2005: 12th ASCOBANS meeting

16th Biennial Conference on the Biology of Marine Mammals, San Diego, USA

19th European Cetacean Society Conference in La Rochelle, France

2004: 11th ASCOBANS meeting

18th European Cetacean Society Conference in Kolmarden, Sweden

2003: 10th ASCOBANS meeting

17th European Cetacean Society Conference in Las Palmas (Gran Canaria), Spain

2002: 9th ASCOBANS meeting

Harbour seal

Wilson RP, Liebsch N, Davies IM, Quintana F, Weimerskirch H, Storch S, Lucke K, Siebert U, Zankl S, Müller G, Zimmer I, Scolaro A, Campagna C, Plötz J, Bornemann H, Teilmann J, McMahon CR (2007). All at sea with animal tracks; methodological and analytical solutions for the analysis of movement. Deep Sea Research II 54:193-210

Adelung D, Kierspel MAM, Liebsch N, Müller G, Wilson RP (2006). Distribution of harbour seals in the German Bight in relation to offshore wind power plants. In: Köller J, Köppel J, Peters W (Eds.). Offshore wind energy. Research on environmental impacts. Springer Verlag, Berlin, pp. 65-75.

Liebsch N (2006). Hankering back to ancestral pasts: constraints on two pinnipeds, *Phoca vitulina & Leptonychotes weddellii*, foraging from central place. Dissertation, University of Kiel

Liebsch N, Wilson RP, Adelung D (2006). Utilisation of time and space by Harbour seals (/*Phoca vitulina vitulina*/) determined by new remote sensing methods. In: von Nordheim H, Boedeker D, Krause (Eds.): Progress in Marine Conservation in Europe – Natura 2000 sites in german offshore waters, Springer, pp. 179-188

Tougaard J, Tougaard S, Jensen RC, Jensen T, Teilmann J, Adelung D, Liebsch N, Müller G (2006). Harbour seals at Horns Reef before, during and after construction of Horns Reef Offshore Wind Farm. Final Report to Vattenfall A/S. Biological papers from the Fisheries and Maritime Museum, No. 5, Esbjerg, Denmark. Available at www.hornsrev.dk

Kiess N (2004). Sichtbarkeit auftauchender mariner Warmblüter in Abhängigkeit von exogenen Faktoren. Diplomathesis, University of Kiel

Trost F (2004). Analyse des Schwimm- und Tauchverhaltens zur Korrektur von Flugzählungen auf See. Diplomathesis, University of Kiel

Wilson RP, Liebsch N (2003). Up-beat motion in swinging limbs: new insights into assessing movement in free-living aquatic vertebrates. Marine Biology 142:537-547

Apart of these publications in journals and monographies, MINOS research network contributes with posters, expert papers and talks to the following prominent international conferences:

2007: 21st European Cetacean Society Conference in San Sebastian, Spain

2006: 20th European Cetacean Society Conference in Gydnia, Poland

2005: 16th Biennial Conference on the Biology of Marine Mammals, San Diego, USA

Seabirds

Markones N (2007). Habitat selection of seabirds in a highly dynamic coastal sea: temporal variation and influence of hydrographic features. Dissertation, University of Kiel

Schwemmer P (2007). Habitat use of the coastal zone of the German North Sea by surface-feeding seabirds. Dissertation, University of Kiel

Garthe S, Flore B-O (2007). Population trend over 100 years and conservation needs of breeding sandwich terns (*Sterna sandvicensis*) at the German North Sea coast. Journal of Ornithology 148: 215-227

Mendel B, Sonntag N, Garthe S (2007). GIS-gestützte Analyse zur Verbreitung und Habitatwahl ausgewählter Seevogelarten in der Ostsee. In: Traub K-P, Kohlus J (Eds.): Geoinformationen für die Küstenzone. Beiträge des 1. Hamburger Symposiums zur Küstenzone. Herbert Wichmann Verlag, Heidelberg, pp. 196-206

Bellebaum J, Garthe S, Kube J, Nehls HW, Schulz A, Skov H (2006). Wasservögel im Küstenmeer Mecklenburg-Vorpommerns: ein Überblick zu Bestandssituation, Gefährdungen und Abgrenzungen neuer Vogelschutzgebiete. Berichte zum Vogelschutz 43:31-47

Dierschke V, Garthe S (2006). Literature review of offshore wind farms with regards to seabirds. BfN-Skripten 186:131-198

Dierschke V, Garthe S, Mendel B (2006). Possible conflicts between offshore wind farms and seabirds in the German sectors of North Sea and Baltic Sea. In: Köller J, Köppel H, Peters W (Eds): Offshore wind energy. Research on environmental impacts. Springer, pp. 121-143

Garthe, S. (2006). Identification of areas of seabird concentrations in the German North Sea and Baltic Sea using aerial and ship-based surveys: In: von Nordheim H, Boedeker D, Krause JC (Eds.): Progress in marine conservation in Europe - Natura 2000 sites in German offshore waters. Springer, pp. 225-238

Schwemmer P, Garthe S (2006). Spatial patterns in at-sea behaviour during spring migration by Little Gulls (*Larus minutus*) in the south-eastern North Sea. Journal of Ornithology 147:354-366

Sonntag N, Mendel B, Garthe S (2006). Die Verbreitung von See- und Wasservögeln in der deutschen Ostsee im Jahresverlauf. Vogelwarte 44:81-112

Garthe S, Schwemmer P (2005). Seabirds at Sea-Untersuchungen in den deutschen Meeresgebieten. Vogelwelt 126:67-74

Schwemmer P, Garthe S (2005). At-sea distribution and behaviour of a surface-feeding seabird, the lesser black-backed gull *Larus fuscus*, and its association with different prey. Marine Ecology Progress Series 285:245-258

Dierschke V, Garthe S, Markones N (2004). Aktionsradien Helgoländer Dreizehenmöwen *Rissa tridactyla* und Trottellummen *Uria aalge* während der Aufzuchtphase. Vogelwelt 125:11-19

Garthe S, Hüppop O (2004). Scaling possible adverse effects of marine wind farms on seabirds: developing and applying a vulnerability index. Journal of Applied Ecology 41:724-734

Sonntag N, Engelhard O, Garthe S (2004). Sommer- und Mauservorkommen von Trauer- und Samtenten (*Melanitta nigra* und *M. fusca*) auf der Oderbank. Vogelwelt 125:77-82

Dierschke V, Hüppop O, Garthe S (2003). Populationsbiologische Schwellen der Unzulässigkeit für Beeinträchtigungen der Meeresumwelt am Beispiel der in der deutschen Nord- und Ostsee vorkommenden Vogelarten. Seevögel 24:61-72

Exo K-M, Hüppop O, Garthe S (2003). Birds and offshore wind farms: a hot topic in marine ecology. Wader Study Group Bulletin 100:50-53

Garthe S (2003). Verteilungsmuster und Bestände von Seevögeln in der Ausschließlichen Wirtschaftszone (AWZ) der deutschen Nord- und Ostsee und Fachvorschläge für EU-Vogelschutzgebiete. Berichte zum Vogelschutz 40:15-56

Garthe S, Krüger T, Kubetzki U, Weichler T (2003). Monitoring von Seevögeln auf See: Gegenwärtiger Stand und Perspektiven. In: Anon.: Vogelmonitoring in Deutschland. Berichte des Landesamtes für Umweltschutz Sachsen-Anhalt, Sonderheft 1/2003:62-64

Garthe S, Ullrich N, Weichler T, Dierschke V, Kubetzki U, Kotzerka J, Krüger T, Sonntag N, Helbig AJ (2003). See- und Wasservögel der deutschen Ostsee - Verbreitung, Gefährdung und Schutz. Bundesamt für Naturschutz, Bonn, pp. 170

Hüppop O, Exo K-M, Garthe S (2002): Empfehlungen für projektbezogene Untersuchungen möglicher bau- und betriebsbedingter Auswirkungen von Offshore-Windenergieanlagen auf Vögel. Berichte zum Vogelschutz 39:75-94

Exo K-M, Hüppop O, Garthe S (2002). Offshore-Windenergieanlagen und Vogelschutz. Seevögel 23:83-95

In addition to these publications in journals and monographies, MINOS research network contributes with posters, expert papers and talks to the following prominent international conferences:

2007: 31st. Waterbird Society Annual Meeting in Barcelona, Spain

42nd European Marine Biology Symposium in Kiel, Germany

Wader Study Group in La Rochelle, France

34nd Annual Meeting of the Pacific Seabird Group in Asilomar, Californien, USA

2006: ICES Annual Science Conference in Maastricht, Netherlands

24th International Ornithological Congress in Hamburg, Germany

9th International Conference of the Atlantic Seabird Group in Aberdeen, Scotland

2005: 32nd Annual Meeting of the Pacific Seabird Group in Portland, Oregon, USA

Common meeting of the Waterbird Society and the Pacific Seabird Group in Portland, Oregon, USA

Workshop on Climate & Seabird populations; University of York; York, Great Britain

11th International Scientific Wadden Sea Symposium; Esbjerg, Denmark

8th International Conference of the Atlantic Seabird Group in Aberdeen, Scotland

All scientists working in the MINOS Research Network contributed to the numerous interim reports and to the final reports of MINOS and MINOS+. At this place, we would like to thank them all very much for their engagement and their successful work. This refers also to their contribution to this book. It was a pleasure to work with you!

List of authors

Dieter Adelung
Leibniz Institute of Marine Sciences
at the University of Kiel
Düsternbrooker Weg 20
24105 Kiel
Germany
www.ifm-geomar.de

Sven Adler
Research & Technology Centre Westcoast,
University of Kiel
Hafentörn 1
25761 Büsum
Germany

Harald Benke
Deutsches Meeresmuseum Stralsund
Katharinenberg 14-20
18439 Stralsund
Germany
www.meeresmuseum.de

Marie-Anne Blanchet
Fjord- & Bælt Centre Kerteminde
Margrethes Plads 1
5300 Kerteminde
Denmark

Michael Dähne
Deutsches Meeresmuseum Stralsund
Katharinenberg 14-20
18439 Stralsund
Germany

Volker Dierschke
Research & Technology Centre Westcoast,
University of Kiel
Hafentörn 1
25761 Büsum
Germany

Jörg Driver
Boßelweg 10
25764 Reinsbüttel
Germany

Kai Eskildsen
Administration of the National Park Wadden Sea of Schleswig-Holstein
Schlossgarten 1
25832 Tönning
Germany
www.minos-info.org (english)
www.minos-info.de (german)

Stefan Garthe
Research & Technology Centre Westcoast,
University of Kiel
Hafentörn 1
25761 Büsum
Germany
www.uni-kiel.de/ftzwest/ag7/

Anita Gilles
Research & Technology Centre Westcoast,
University of Kiel
Hafentörn 1
25761 Büsum
Germany

Nils Guse
Research & Technology Centre Westcoast,
University of Kiel
Hafentörn 1
25761 Büsum
Germany

Helena Herr
Research & Technology Centre Westcoast,
University of Kiel
Hafentörn 1
25761 Büsum
Germany

Christopher G. Honnef
Deutsches Meeresmuseum Stralsund
Katharinenberg 14-20
18439 Stralsund
Germany

Annette Kilian
Tiergarten Nürnberg
Am Tiergarten 30
90480 Nürnberg
Germany

Kristina Lehnert
Research & Technology Centre Westcoast,
University of Kiel
Hafentörn 1
25761 Büsum
Germany

Paul A. Lepper
Applied Signal Processing Group
Dept. Electronic & Electronical
Engineering
Lougborough University
Lougborough, UK

Nikolai Liebsch
Institute of Environmental Sustainability,
Biological Sciences
University of Wales Swansea
Singleton Park, SA2 8PP
Swansea, UK

Klaus Lucke
Research & Technology Centre Westcoast,
University of Kiel
Hafentörn 1
25761 Büsum
Germany

Nele Markones
Research & Technology Centre Westcoast,
University of Kiel
Hafentörn 1
25761 Büsum
Germany

Anja Meding
Deutsches Meeresmuseum Stralsund
Katharinenberg 14-20
18439 Stralsund
Germany

Gabriele Müller
Leibniz Institute of Marine Sciences at the
University of Kiel
Düsternbrooker Weg 20
24105 Kiel
Germany

Tanja Rosenberger
Seehundstation Friedrichskoog
- Seal Centre Friedrichskoog -
An der Seeschleuse 4
25718 Friedrichskoog
Germany

Jacob Rye
Research & Technology Centre Westcoast,
University of Kiel
Hafentörn 1
25761 Büsum
Germany

Meike Scheidat
Wageningen IMARES
Institute of Marine Resources and
Ecosystem Studies,
Dept. Ecology
Postbus 167
1790 AD Den Burg
The Netherlands

Ursula Siebert
Research & Technology Centre Westcoast,
University of Kiel
Hafentörn 1
25761 Büsum
Germany
www.uni-kiel.de/ftzwest/ag7/

Nicole Sonntag
Research & Technology Centre Westcoast,
University of Kiel
Hafentörn 1
25761 Büsum
Germany

Janne Sundermeyer
Research & Technology Centre Westcoast,
University of Kiel
Hafentörn 1
25761 Büsum
Germany

Ursula Katharina Verfuß
Deutsches Meeresmuseum Stralsund
Katharinenberg 14-20
18439 Stralsund
Germany

Katrin Wollny-Goerke
Kakenhaner Weg 170
22397 Hamburg
www.meeresmedien.de

List of photos

Title:	
Background	Katrin Wollny-Goerke
Harbour porpoise	Klaus Lucke, Research & Technology Centre Westcoast / Fjord- & Bælt Centre Kerteminde
Harbour seals	Katrin Wollny-Goerke
Black-legged Kittiwake	Sven-Erik Arndt
Scientists on board at survey	Nicole Sonntag
Windpark Horn´s Rev	Ansgar Diederichs
Chapter 1	
Wind farm Horn´s Rev	Ansgar Diederichs
Chapter 3	
TPOD	Ursula Verfuß
Calibration set-up	Ursula Verfuß
Chapter 4	
Airgun	Klaus Lucke
Chapter 5	
Dead-reckoner	Gabriele Müller
Harbour seal with dead-reckoner	Gabriele Müller
Chapter 8	
Great Cormorant feeding flock	Nicole Sonntag
Species - Excursus:	
Harbour porpoise 1	Harald Benke
Harbour porpoise 2	Tanja Rosenberger, Seehundstation Friedrichskoog / Fjord- & Bælt Centre Kerteminde
Harbour seal 1	Katrin Wollny-Goerke
Harbour seal 2	Katrin Wollny-Goerke
Herring Gull	Sven-Erik Arndt
Common Gull	Stefan Garthe
Black-legged Kittiwake	Sven-Erik Arndt
Lesser Black-backed Gull	Stefan Garthe
Black-headed Gull 1 + 2	Sven-Erik Arndt
Common Guillemot	Sven-Erik Arndt
Great Cormorant	Hans-Georg Arndt